The Changing Rol

Active Le...
Enterprise ...

D1147949

ꓛYL'SH CAMP'ꓔS

The Changing Role of Specialist and Trade Contractors

Colin Gray
University of Reading

Roger Flanagan
University of Reading

Chartered Institute of Building
Ascot

Contents

Acknowledgements

We are indebted to four groups: the Chartered Institute of Building, the Institute's Research Steering Committee, our research team, and the vast number of contractors and sub-contractors who have made our task so interesting and enjoyable. To mention everyone would be impossible, but to those not mentioned we extend our thanks.

This is the largest single research study commissioned by the Chartered Institute of Building. We submitted a proposal for this study as it was evident that there needed to be a better understanding of the role that the sub-contractor was being asked to play. The Institute recognises that it has a role in helping to formulate strategy for the whole of the construction industry and through the developing programme, which is the responsibility of the Institute's Research Committee, funded the study. We thank the members of the Institute's Research Committee for their confidence. Our special thanks is extended to Peter Harlow of the Institute for his adept and able work behind the scenes in order to smooth our path.

A steering committee was established under the Chairmanship of Peter Titmus, and although there have been one or two changes during the study period their guidance has been most valuable and enthusiastic. The full membership is:

P.D.I. Titmus	Fairclough Building Ltd. (Chairman)
P.A. Harlow	The Chartered Institute of Building
G.P. Cottrell	The Electrical Contractors' Association
H.R Howard	The Federation of Building Specialist Contractors

P.J. Long	McKenna and Co.
A Massey	The APC International Group
R O'Rourke	R O'Rourke & Sons
J Price	Heating and Ventilating Contractors' Association (HVCA)
J Leeder	HVCA
E.H. Potter	Higgs and Hill Construction Holdings Ltd.

We have been fortunate in being supported by a very good research team and special thanks must be given to: He Zhi, John Jewell, Carol Jewell, Kevin Mole, Ren Hong, Ho Kwan Yu, Mark Eustace. Also to Tim Cornick and Dr Brian Atkin for major input into the sections on Quality Management and Research and Development respectively.

C Gray
R Flanagan

Reading, September 1989

Foreword

I am sure that this research into the sub-contracting sector of the construction industry is the most comprehensive analysis of its kind.

It is clear from the research that the nature of sub-contracting is such as to defy precise definition; there are no absolutes or established norms. The research team has nonetheless, with creditable clarity provided us all with an insight into the structure, processes and relationships, that relate to this diverse industry.

One cannot fail to recognise the strength of opinion voiced by all parties to the construction process when when the 'problems' of the industry are discussed. The notion that individual perceptions are open to challenge, is not the point. It is the very fact that rightly or wrongly many such opinions are universally held that begs our examination.

Our research team, very ably lead by Colin Gray and Roger Flanagan at the University of Reading, have successfully initiated the debate. They have provided us with a detailed analysis of the issues and have invited us to consider the future strategy. On behalf of the Chartered Institute of Building, I am indebted to them for the quality of their deliberations.

The Chartered Institute of Building is also indebted to the members of the Steering Committee who have generously given of their time and who have provided a substantial level of expertise and informed comment at the various draft report stages. Our grateful thanks also extend to Peter Harlow for his efficient and patient administration.

As our industry begins to face up to the challenges of the 1990s, the gathering pace of the changes forces us to examine our existing customs and practices. It motivates us to consider the shape of things to come in a wider European context. Clearly our sub-contracting industry must be fit and ready to meet this challenge. The Chartered Institute of Building is to be commended for commissioning such an important study which is part of the continuing programme of the Institute's Research Committee.

P.D.I. Titmus MSc. MCIOB
Chairman of the Steering Committee

Introduction

The construction industry in 1989 is very different from the industry of ten years ago. Despite frequent criticism the industry is leaner, fitter, and given the opportunity, is now able to build faster and to higher standards. The 1980s have, therefore, seen unprecedented change in the UK construction industry which has accelerated, predominantly due to political and economic pressure. Much of the pressure has come from the industry's clients, but there has been a significant restructuring of the industry due to sophisticated financial management and the specialization of technology and its application. There has been a significant weakening of both the architect's design role and capability and the contractor's direct production role. The resolution of both of these weaknesses has been found in the specialist contracting sector of the industry. The change has been so marked that over 90% of construction activity is now sub-contracted on the majority of contracts (Faster Building for Commerce, 1988).

The performance of the sub-contractor is now critical to project success, yet sub-contracting does not have a glamorous image. This is due mainly to its past association with the traumas of lump labour, tax evasion, cowboy operators and the large number of bankruptcies which litter the scene of the small sub-contractor. Labour only sub-contracting has received the invective of the unions and it is only recently that attitudes have started to change because the predominant proportion of the workforce is now employed by sub-contractors. It is now a high risk business with the majority of the risks, financial, technical and labour, borne by the sub-contractors. Contractors and designers have increasingly off loaded the risk onto the sub-contractors with a rising number of onerous contracts and obligations.

The background to the research

The research was undertaken by a group of researchers who conducted interviews with clients, architects, engineers, quantity surveyors, general contractors, specialist contractors, trade associations and government departments. Use was also made of questionnaires to obtain quantitative information.

The views of the industry

With a subject as complex as sub-contracting there were widely differing views. At the outset we feel it is worth reproducing some of the opinions voiced at the interviews. They are quoted below as showing the viewpoints of the client, the design consultants, the contractors and the specialist contractor. The quotations have been condensed, but not substantially edited. They help to show that each group sees the situation from their own perspective. Unfortunately, they also show that there are serious misgivings about the performance sub-contractors by all the parties.

The client's viewpoint

❑ "The industry has a very cavalier attitude to after sales care for its clients. I am interested in getting the product on price and on time. When a defect occurs after completion nobody wants to help with the problem and I am left having to join parties in legal action. The worst people to get hold of are the specialist sub-contractors.

❑ The only certainty after birth is death, or so I thought, but I have discovered another on my projects; the submission of claims for loss and expense."

The design consultant's viewpoint

❑ "The specialists don't understand the problems of aesthetics and difficulties of aesthetics and design.

❑ The specialists have the practical skills, which are useful at the design stage, but it is difficult to take account of all their views at the design stage without committing ourselves to them and their price, and if they are not in competition the price will be too high.

❏ We would love to have an input from the specialists at the design stage, but not on a fee paying basis.

❏ In the competitive environment we work in, whilst we try to have a good relationship with the specialists, most projects are awarded to the lowest bidder.

❏ Technology and specialism are developing so fast that we have to rely on shop drawings produced by the specialists both at pre and post construction phase.

❏ Low professional fees are forcing designers to concentrate more on design and leaving the specialists and contractors to sort out the nitty-gritty construction details.

❏ Clients are becoming more litigious and the problems of latent defects are becoming more serious, when things go wrong the designer is always implicated.

❏ There is little time to prevent the client suing you if you gave 'approval' of the specialist's drawings negligently, and when things go badly wrong the smaller specialists go into liquidation.

❏ Whilst many specialists undertake an element of the design work, very few carry a design liability professional indemnity insurance policy. Many specialists are not even aware they need a separate policy to their 'all risks' construction policies.

❏ There are always communication problems when we have to deal with a local consultant whom the specialist has commissioned to carry out a small design commission for their work package.

❏ Chairing site meetings is always a nightmare because the specialists argue amongst themselves about interface problems, nobody wants to own a problem. It is always the architect causing the delay.

❏ The larger specialists, like curtain walling and lift manufacturers, are developing a good commercial awareness; they are very professional and very reliable even when things go wrong. The downside is that they want to impose their own contractual conditions upon us rather than accepting the contract conditions in the standard forms of contract.

❏ Specialists did not need to read the small print in the conditions of contract in the past, but now they are becoming more contractually aware and the conditions are now read in detail to determine the problems and opportunities, whereas before the contract was just filed and a working relationship developed

❏ The trend towards collateral warranties required by the client is good in theory but the practical difficulties have left the system in a mess."

The contractor's viewpoint

❏ "Although an increasing proportion of work on site is now undertaken by specialists, the problems of getting the project constructed have increased in complexity because of the interface problems.

❏ Not all specialist and trade contractors have good management skills and many firms need a lot of mothering to ensure that they keep to programme and achieve the quality required.

❏ When things go badly wrong many of the smaller specialists go into liquidation and cease trading, only to appear under another company name. As a result the contractor is still ultimately responsible to ensure that any latent defects in the work are remedied.

❏ Whilst all specialists are interested in the quality of their own workmanship and products, we are interested in the quality of the completed project. Too frequently the specialists do not care about the other trades and we are left to sort out the re-work disputes and the cost of the return visits to site.

❏ We are responsible for the overall safety of the site and providing and checking a safe system of work, but the safety systems have to be constantly monitored because of a few poorly trained workers working for specialists who are only interested in achieving the bonus payments.

❏ We are blamed by the specialist trade contractors for imposing unreasonable and onerous conditions on them, but they fail to realize that many of these are imposed by the clients and their professional advisers who extensively amend or use non-standard forms of building contract

❑ Getting paid is the biggest difficulty facing the industry, we are blamed for tardiness in the valuation and payment of money yet it is the clients that are the most tardy especially where the funding arrangements are complex and there are many parties involved.

❑ Whilst we try to maintain a good working relationship and to ensure repeat business with approved specialists, the competitive tendering system means we must see price as the overriding feature and frequently this means an inability to offer repeat business."

The specialist contractor's viewpoint

❑ "More of the financial and technical risk is being passed to the specialists without the appropriate financial rewards.

❑ Specialist contractors are everybody's 'Aunt Sally'.

❑ The specialists and trade contractors have very little influence whilst onerous contract conditions are imposed unfairly, such as 'pay when paid' clauses, unqualified rights of set off and deletion of the right to appoint an adjudicator to settle disputes.

❑ Quality assurance is being passed down the line to the specialists with all the paper-work and aggravation and few of the benefits, Quality schemes should start from the workface and go upwards.

❑ Quality is about the attitude of, and communication with the workers, not more regulations and checks.

❑ There has been a decline in training as a result of the 'stop-go' policies in the past which resulted in a fluctuating work load; in order to stay profitable in the downturn we had to reduce the workforce and the number of trainees.

❑ The shift from directly employed to self employed operatives is still continuing. There is now a lack of craft skills.In the future the situation could continue to deteriorate due to the lack of trained craftsmen.

❑ Requesting performance bonds for relatively small packages is a trend that is increasing.

❑ If only the architects would consult us at the design stage a lot of the design details that are impractical could be avoided. Lip service is paid to buildability, but in reality the sequence of operations is not a high priority to a designer.

❑ Clients change their minds and architects issue variation orders for work that had not been adequately designed at the pre-contract stage, without thinking of the implications for the sub-contractor. As a small company we might be working on five sites and balancing our resources across the projects. Even minor changes can play havoc with this.

❑ Clients expect the sub-contractor to respond at will. This is possible only if the design is complete and we know what we are doing, rather than working with a design which is continuously evolving as we go along."

Each set of viewpoints gives the impression that everything is difficult and it is nearly always somebody else's fault. They demonstrate a lack of understanding which we all have for each other's problems. If we are to be honest the true position is:

❑ The client is interested in the product not the process.

❑ The contractor is interested in how the process can deliver the product.

❑ The sub-contractor is interested in his product and suffers the process.

❑ The supplier is interested in getting his product out of the factory door, not in the rest of the process.

Structure of the report

This study has attempted to look at the past, present and future of sub-contracting in the UK construction industry. In order to provide a better basis for the future the many points raised above can only be answered by analysing the historical development both of sub-contracting and the aspects of the industry which impinge onto the spectrum of sub-contracting. In Chapter 1 the factors which have shaped the current state of sub-contracting are examined. They are economic, political, technological and organizational. The industry, as a consequence, is a very complex one where trying to produce good quality work economically and to a schedule is a difficult task, as shown in the case study in Chapter 2. Sub-contracting is an integral part of the whole process of construction and yet the complexities and the ambitions of the various parties in this process are not well understood. In Chapter 3 a model of the process is developed which will be used throughout the study to explain relationships, explore some of the issues, and to describe the problems regarding the definition of responsibilities. By far the largest section in this report deals with formal contractual relationships and the problems that seem to arise. In order to clarify the issues the basic process model is used in Chapter 4 to examine the responsibilities of the designers, sub-contractors and contractors under the principles of the various forms of contract. Many of the contracts are drawn up with an imperfect knowledge of the processes of construction, let alone the role that the sub-contractor is to play.in those processes In many cases the contracts are attempts to manipulate the processes of construction but fail. The integral role that the specialist trade contractor now plays within the construction process is both complex and important. Some of the biggest problems occur over the manipulation of contracts and these issues are examined in Chapter 5. The most complex area in which the sub-contracting industry is now involved is the design process. Chapter 6 discusses thoroughly the issues of liability and warranty.

Sub-contractors are increasingly being presured into taking greater responsibility for the total package of work. The aim of this is to maximize the production process, but there are many conflicting demands, as described in Chapter 7, before this can be achieved. The current major issues which face the sub-contracting industry are then examined in the next three chapters. In Chapter 8 perhaps the most topical problem currently facing the industry is examined, that of achieving a quality product. The issue is complex because total product quality is a matter which concerns all, yet very few see the total product. The way to quality is through the operatives and there are close links to productivity, but the industry has had a cavalier attitude to safety and its consequent demotivation of the workforce, as shown in Chapter 9, The demographic trends are not encouraging as shown in Chapter 10, where the whole issue of recruitment and training for both operatives, site supervision and management is discussed.

The final section looks to the future. In Chapter 11 the research and development needs of sub-contractors are examined. It is an area which attracts a lot of criticism, much of which contains a grain of truth, and there is need to change fundamental attitudes to business if R & D is to be the driving force that many think is necessary. Much of the pressure to change is coming from the following overseas sources: companies which appear to be research led; contracting systems based upon a different concept of sub-contracting, and systems with differing relationship structures. These are explored in Chapter 12 by reference back to the construction process model developed in Chapter 3.

Much of the report attempts to explain the reasons why the three parties to the construction process seem to have such disparaging views of the role and activity of sub-contractors. Whilst there are other process models overseas, and some of the ideas and practices are transferable, the UK industry must seek its own solution. If it does not, clients and contractors will continue to drive the sub-contractors in ways which may not be in their best interests or those of the industry. A blue-print for the future is developed in Chapter 13.

Terminology

Various terms are used to describe the contractors who undertake specialist work:

Sub-contractor
Sub-sub-contractor
Specialist contractor
Works contractor
Trade contractor
Specialist trade contractor
Specialty contractor
Work package contractor
Labour only sub-contractor

Our aim is not to argue the merits or demerits of the best terminology. Contractual terminology refers to the sub-contractor or works contractor, whilst trade associations prefer to use specialist contractor. Throughout the text All of these terms have been used interchangeably where appropriate to refer to the different aspects of sub-contracting.

Similarly, the terms contractor, general contractor, prime contractor and main contractor have been used to refer to the building contractor where appropriate.

We have used the term construction industry which includes building and civil engineering and othe enginering work. However all of the discussion and references to the contract relate to building work.

1

The development and organization of sub-contracting

There has been no single factor which has accelerated the development of sub-contracting. It has resulted from a gradual squeezing of employment opportunities as the construction industry has responded to the variations in workload over the last twenty years. A variety of pressures, from depression to boom, from high levels of employment to unemployment, from traditional technology to high technology, from predominance of public sector investment to a predominance of private sector investment and the need to build faster, have all played their part in changing the construction industry. The emergence of the sub-contracted labour base of the industry has been a significant response to the volatility of these changes.

Volatile workload

Construction is the one of the few industries in which the work place is variable and the worker has to be a mobile unit. The cyclical nature of the industry as a whole and the more volatile local changes within these cycles led to an industry structure of large organizations and contractors that employed a core workforce, which was supplemented with locally recruited workers as required. This worked well when the industry was growing (1955 - 1968), but since the levelling in demand (1968 - 1973) and the sharp fall since then (see figure 1.1), there has been a reduction in the core workforce and the itinerant workforce (see figure 1.2).

Yesterday	Today
Clients prepared to accept what the industry told them was the accepted practice.	Clients are more discerning and demanding of better value for money and built faster.
A dominant public sector client with over 55% of the total workload funded by the public sector.	The emergence of a dominant private sector with over 60% of the workload funded by the private sector.
Volatile workload with stop go policies using the construction industry work load as a regulator for economic activity.	A less volatile workload in the industry with the construction workload more subject to market forces, such as interest rates and government fiscal policy. Overall less government interference.
The success of any project is dependent upon the attitude of the parties.	The success of any project is still dependent on the attitude of the parties.
No significant European perspective for the industry.	A European perspective that offers both a threat and an opportunity.
Low profit margins on contracting with little capital required to enter the industry.	Profit margins on contracting are still low but more diversification has taken place with the major contractors being involved as developers, materials suppliers and speculative developers.
Lump sum price, either fixed price or fluctuating price being the most usual.	Lump sum price still the most usual. Innovative methods of procurement now emerge.
No competition from overseas companies in the UK market.	Increasing overseas competition from material suppliers, specialist contractors, with UK contractors being taken over by overseas conglomerates.
Litigation to resolve disputes seen as a last resort.	Litigation and 'seeking the guilty' becoming more accepted.

Yesterday	Today
National and regional wage negotiations undertaken in wages and conditions committees and seen as having a real influence on wage levels for site operatives.	Wage negotiations still undertaken in committees but of less significance to site operatives who generally earn above the basic wage rate. Greater incentives to operatives for increased production.
Quality of workmanship and materials seen as important.	Quality of workmanship and materials seen as a major issue with quality assurance being regulated by BS 5750
Clerk of Works seen as a check to ensure proper quality standards	Few Clerk of Works used on sites.
Some off- site pre-fabrication of components.	Considerable off-site pre-fabrication of components.
Design rarely complete at the tender stage.	Design still rarely complete at the tender stages.
Fee scales for professional services set by the professional institutions and associations.	Fee competition for professional services now accepted practice.
Professional indemnity insurance for consultants, necessary but not expensive.	Professional indemnity insurance for consultants now imperative and very expensive.
Nomination of sub-contractors by the architect for specialist work is very common.	Substantial reduction in nomination of sub-contractors, move towards named sub-contractors.
Specialist sub-contractors undertaking some design input to a project.	Increasing design input required from sub-contractors.
Information technology not having much impact on construction sites.	Information technology seen as being vitally important for the future.

Yesterday	Today
Some sub-contracting and labour only workers on sites.	Extensive use of sub-contracting and labour only workers, particularly in the south east.
The role of the specialist sub-contractor is important.	Sub-contracting becoming more specialized and is now of vital importance to the Industry.
Apprenticeships and training taken seriously by contractors. The only route to become recognized as a craftsman was by apprenticeship.	Fewer apprenticeships and reduction in the amount of training undertaken by contractors because of the increasing use of sub-contracting.
Craft skills seen as being the most important for tradesmen.	Craft skills still important but product specific skills and multi-skilling for operatives seen as a growing need.
Apprenticeship based upon time-serving as the sole requirement for craft recognition.	Skills testing of apprentices which takes account of the fact that people learn at different speeds and have different levels of ability and varying degrees of experience.
The centralising of training for the construction industry is seen as a major break-through with the establishment of the Construction Industry Training Board (CITB).	CITB under attack and not seen as adequately satisfying all the various skill training needs.
Real trade union influence and power on the larger construction sites.	A reduction in the trade influence on construction sites.
Safety of operatives on site is a problem.	Safety of operatives on site is still a problem.
General foremen and site managers mainly recruited with craft background.	General foremen and site managers now recruited from craft background and University or Polytechnic graduates.

Yesterday	Today
Construction contracts filed but not frequently read unless a dispute occurs.	Construction contracts scrutinised in detail, often by specialists lawyers.
Less awareness of risk.	Passing the risk to another party is seen as a necessity.
Construction claims for reimbursement of loss expense seen as the only way to earn profit during slump periods.	Construction claims for reimbursement of loss and expense becoming a scientific exercise with the emergence of construction disputes resolution consultants. Overall claims are becoming fewer but they are becoming more sophisticated.
Collateral warranties required by the clients very rare for projects.	Collateral warranties required by everybody, becoming prolific and very complicated for projects.
Performance bonds not usually requires from contractors and sub-contractors	Performance bonds becoming requires more frequently.
Most work undertaken using standard forms of contract.	Work undertaken on standard forms of contract that are often amended to suit the needs of the employer.
Many contracts undertaken based on trust and understanding between the parties.	Contractual obligations based solely on written agreement with the contract conditions attempting to cater for most eventualities.
Some onerous conditions inserted into contracts between employers and contractors, and between contractors and specialist contractors.	Many onerous conditions being inserted into contracts.
Poorly organized small sub-contracting companies.	Better organized and larger specialist sub-contractors beginning to emerge.

There has also been significant changes in the composition of the workload with a dramatic decrease from 51.6% to 21.06% in public sector work (figure 1.3) as well as a very large growth in repair and maintenance (figure 1.4).

Contractors have increasingly had to seek work further and further from their base and so have had to recruit the workforce locally, as the cost of using their own workforce has been too high. Most importantly the employment legislation over the past twenty years, which has imposed financial burdens on employers, such as redundancy payments, employers' national insurance contributions, sickness benefits and holidays with pay, have all led to the pressure to reduce fixed costs. Workers also started to object to this volatility in their lifestyle and realized that conventional wage agreements were no compensation. They increasingly chose to become self employed (figures 1.5 and 1.6) and thereby able to set their own rate for the work. This became abused with the 'lump' but clearly set the trend for the growth in sub-contracting.

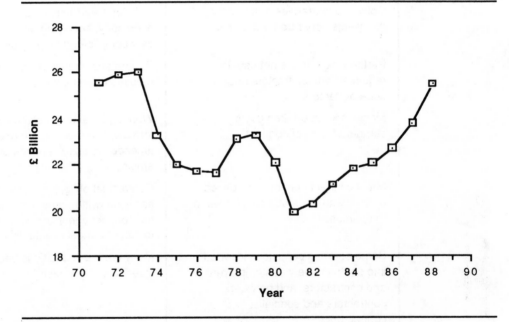

Figure 1.1   Total British Construction Output per annum, 1971-1988 (1980 current prices)
Source: Housing & Construction Statistics, DoE

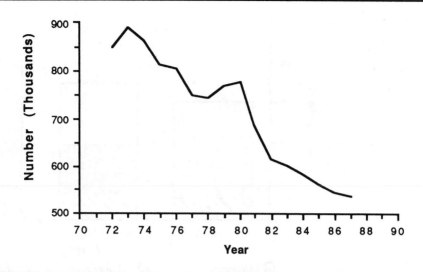

Figure 1.2   Operatives directly employed by private contractors, 1972 - 1987
(Excl. self employed).
Source: Housing & Construction Statistics, DoE

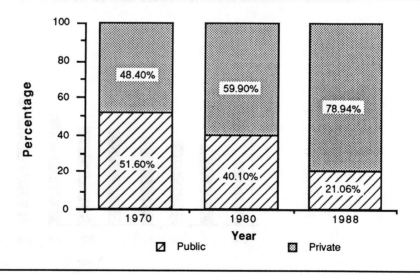

Figure 1.3   Change in proportion of total workload for public sector work
Source: Housing & Construction Statistics, DoE

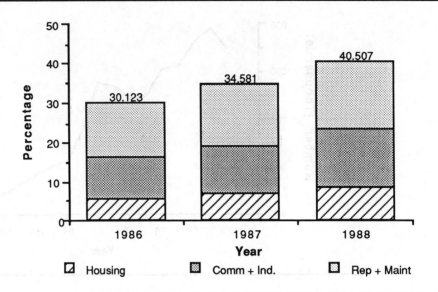

Figure 1.4   Sector workload patterns as a proportion of total workload
Source: Housing & Construction Statistics, DoE

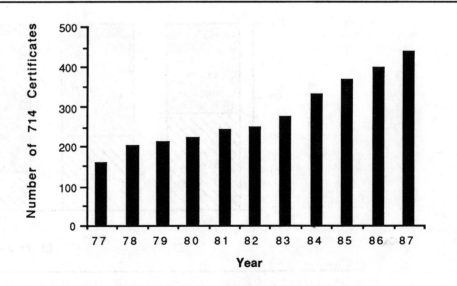

Figure 1.5   Individual holders of 714 Certificates, 1977 - 1987
Source: Income data services, Study 396, October 1987

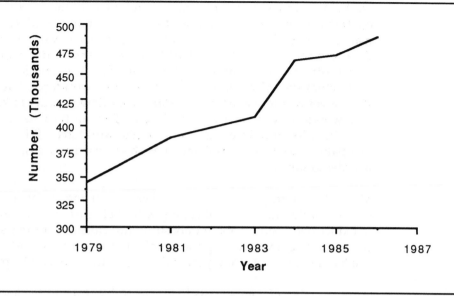

Figure 1.6   Estimates of self-employed, 1979-1987
Source: Department of Employment, Employment Gazette

Technology

The construction site is often considered to be the place where the construction process takes place. This view may have been true at the turn of the century when the traditional crafts of bricklaying, carpentry, tiling, thatching and masonry were used to construct the whole building from the raw materials delivered to site. Since that time however, many factors have caused an increasingly rapid shift away from site based manufacture. Building construction is now largely a process of assembling an enormous variety of components, many of which are made in factories under controlled conditions. The components are still assembled primarily by the traditional craft skills as the jointing techniques have largely remained the same, but there are also new skills and modified traditional skills which have been created by the needs of the new components.

Technology was also starting to develop in the 1960s with a period of enthusiastic experimentation. The country was recovering fast from the restraints of the Second World War and there was a new political will to provide housing for the population. The demand outstripped the ability to supply and technology was used to develop new methods of building. Many of these were developed by specialist manufacturers who also undertook the site work. As projects started to contain many more of these specialist components, contractors increasingly found their work nominated to these specialist manufacturers. The contractor could no longer even rely on a known work content in a project and so the volatility of his labour demand rose with the variability of the technology being used. The specialist suppliers and contractors started to employ the contractor's workforce, but found that they themselves had to maintain a constant work flow in order to keep even a small core workforce economically employed. They in turn started to sub-contract work in order to reduce the risk of an idle workforce. When the recession hit, this pattern accelerated as companies shed their fixed costs as fast as possible in order to survive.

Many sub-contractors let most of their site fixing to labour only operatives because of the difficulties of coping with their variable workload and the need to reduce overheads. Labour only sub-contractors now form a significant part of the workforce, exacerbating the problem of operatives working together on a site having different employers and variable terms of employment.

As the number of components has increased and the technology changed so the knowledge has become specialized. To build one building, therefore, requires the knowledge of many specialists.

The power of contractors

Contractors have evolved into merchant traders (Ball 1988). This process began with sub-contracting and now over 90% of the site work in many areas of the UK is sub-contracted, the process is complete. They are now managers of other organizations and their power lies in their knowledge of the whole process and the fact that they are the purchaser of the parts for the process. Nomination removes a large amount of the discretion that the contractor had in the selection process but, once the choice is made and the contract sealed, the contractor is in control.

The sub-contractor is in a weak position to bargain or negotiate as he is once removed from the decision process between client, designer and contractor. The contractor has absolute control over the terms and conditions as well as payment procedures. This financial control has played a very large part in the relationship between contractors and sub-contractors as contractors have realized that their primary asset is work volume and cash flow. Profit levels have always been very small, but by judicious manipulation of payment cycles, cash flow can be enhanced and working capital reduced or virtually eliminated. This accounting view of contracting has accelerated the use of sub-contracting with the consequential shedding of financial risk by contractors.

Labour relations

The days of rapid hire and fire are long gone. Employment legislation introduced in the 1960s was designed to give the worker increased rights, particularly over discipline and dismissal. Such legislation is difficult to enforce in the construction industry at anytime, with a contractor's requirements varying weekly or even daily due to project requirements. However, in the recessions of the 1970s and early 1980s, keeping a workforce regularly occupied was increasingly difficult and consequently contractors shed labour at every opportunity. It was also an uncomfortable period for labour relations, with tremendous industrial unrest spilling over into construction. The number of days lost to strikes reached a peak in 1972 and contractors sought ways of reducing their risk. The most obvious way was to employ either small companies who did not have a unionized workforce, or individuals who were not union members. So a trend towards removing the risks associated with large, unionized workforces began.

The recession became so deep that workers desperate for work rapidly lost interest in striking and other disruptive activity and the dawning of a new era of legislation introduced by the Thatcher administration progressively curbed the union's freedoms throughout the 1980s. Days lost through strikes fell to an all time low in 1986, at only 9% of the level in 1979, see the graph in figure 1.7. However, there was a subsequent realisation, particularly on the major London contracts, that, whilst they may be large, they were determined not to suffer from the industrial unrest of the big sites in the 1960s. These sites actively encouraged all workers to become members of the union in order to control the union's membership ambitions, control inter union wrangling, ensure compatibility between trades, and provide good welfare facilities.

Sub-contracting today

It is impossible to generalize about sub-contracting because the range of organizations embraced under the umbrella of sub-contracting is so diverse. There is a predominance of small firms, but the range is from one man self employed to large multi-national industrial groups, giving a range of services from labour only to large scale integrated manufacturing and erection. As shown in figure 1.8, there are four main categories ranging from 'fix only' to a combination of manufacturing and design through to the full design, manufacture, supply and fix of complex systems. Fix only would be, for example, a brickwork sub-contractor, whereas supply and fix would be a glazing or painting sub-contractor. At the other extreme including design and manufacture would be the specialist curtain walling sub-contractor. Many companies will offer a combination of these options tailored to the specific project demands. Figure 1.9 illustrates the view of Bennett and Ferry, shown in a range of design contributions with the comment that the sub-contractor can "enter the process at any of the stages shown".

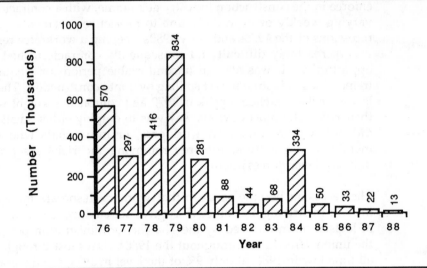

Figure 1.7   Days lost due to industrial disputes
Source: Department of Employment, Employment Gazette

As well as being involved in this wide range of activities, some important features about sub-contracting are:

❑   Not all sub-contractors are engaged in the finished product, for instance the scaffolding sub-contractor provides temporary work, or the plant hirer provides the driver and equipment for another specialist to use.

❑   Many sub-contracting companies are multi-industry, in other words the construction industry is not the only industry that uses their services; for example, for some electrical contractors the construction industry is only 20% of their business.

❑   Many of the sub-contractors will employ their own labour force direct whilst others will sub-let to labour only firms.

❑   The size of the sub-contracting companies ranges from the one-man firm to the multi-national firm such as Redlands.

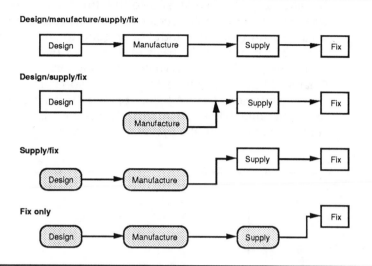

Figure 1.8   Range of sub-contractor contributions

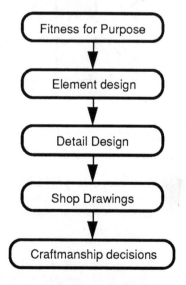

Figure 1.9   Range of specialist's design responsibilities
Source: Specialist Contracting - A review of issues raised in their new role
in building, The Building Centre Trust

In summary, specialist sub-contracting is a very diverse, fragmented, specialized and complex sector of the construction industry. The diverse and volatile nature of "here today and gone tomorrow" sub-contractors is often criticised, mainly on the basis of their poor financial performance. However, given the problems of payment and onerous conditions (to be discussed in Chapter 5) the financial performance of specialist contractors is on a par with that of contractors (figure 1.10). It is to be expected that the contractor's profits are higher due to the cushioning effect of his other interests and the use of sub-contractors to smooth the irregularities of construction work.

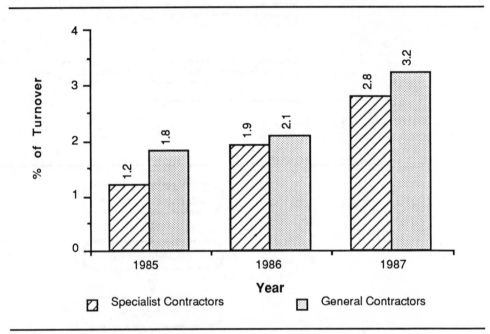

Figure 1.10   Percentage of profit before tax on turnover Specialist contractors and general contractors in the UK (1985-1987)
Source: Key Business Ratios, Third Edition 1989, (Dun and Bradstreet International)

Notes: 1985   9592 Companies for Specialist Contractors
                      7507 Companies for General Contractors
           1986   9535 Companies for Specialist Contractors
                      7947 Companies for General Contractors
           1987   7314 Companies for Specialist Contractors
                      6492 Companies for General Contractors

The average profits shown in figure 1.10 do not indicate the range of performances within the various sub-sectors, where high profit figures in plant hire, excavation and flooring support the other, generally much weaker, sectors (see table 1.1). Within the sub-sectors themselves, there is again a wide variability of performance. Figure 1.11 shows that the 1986-1987 boom improved the overall performance considerably, with everyone in the sample reporting profitable trading. Specific sub-sector performance, such as heating and ventilating contractors indicates that the contractors' profitability has been within a fairly limited range of between 1% and 4% (see figure 1.12). Although this profitability follows the peaks and troughs of the workload cycle there is a general lag of two to three years.

SPECIALISATION	1985	1986	1987	AVERAGE
Plumbing, heating, air conditioning	0.2	0.9	2.9	1.3
Painting, decoration	0.8	1.5	1.9	1.4
Electrical works	0.2	1.8	2.7	1.6
Masonry, stonework	4.3	-0.5	1.4	1.7
Plastering, insulation	-1.6	2.1	3.6	1.4
Tile marble, mosaic work	0.9	3.1	1.2	1.7
Carpenters	-1.7	2.1	2.5	1.0
Floor laying, floorwork	2.4	2.0	4.7	3.0
Roofing, sheet metal work	0.6	2.8	2.3	1.9
Concrete work	1.7	3.3	1.8	2.3
Structural steel erection	1.0	0.8	1.9	1.2
Glass, glazing work	-0.8	1.0	3.2	1.1
Excavating, foundation	4.7	4.6	1.9	3.7
Wrecking, demolition	1.1	1.2	2.2	1.5
Miscellaneous building equip instal.	2.1	0.7	3.4	2.1
Plant hire	5.3	4.9	8.4	6.2
Miscellaneous special building trade	-1.1	0.8	1.3	0.3
Average	1.2	1.9	2.8	2.0
Maximum	5.3	4.9	8.4	6.2
Minimum	-1.7	-0.5	1.2	0.3

Table 1.1   Percentage of profit before tax on turnover. Specialist contractors in the UK (1985-87)
Source: Key Business Ratios, Third Edition 1989, (Dun and Bradstreet International)

Notes: 1985 - 9,592 Companies
1986 - 9,535 Companies
1987 - 7,314 Companies

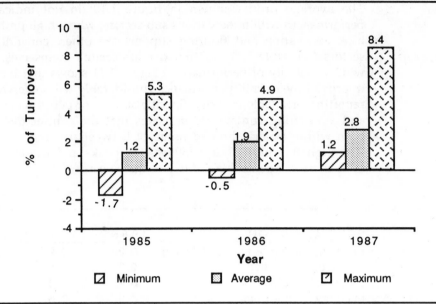

Figure 1.11 Percentage of profit before tax on turnover. Specialist contractors in the UK (1985-87)

Source: Key Business Ratios, Third Edition 1989, (Dun and Bradstreet International)

Notes: 1985 - 9,592 Companies
              1986 - 9,535 Companies
              1987 - 7,314 Companies

As shown in figure 1.13 the rate of growth in 1984 in the number of specialist contractors has been the highest of all contractors since 1977. However, as well as being one of the easiest industries to join it is also one of the easiest in which to fail as shown by the scale of insolvencies in the industry. (figures 1.14 and 1.15) The number of small, self-employed, construction organizations that fail has increased in 1987 (figure 1.14). Whilst the number of construction companies that became insolvent is interesting (figure 1.15), what really matters is the volume of debt which is left by the companies. It has not been possible to separate the number of failures for specialist contractors from the overall data. However, it can be assumed that because the highest proportion of failures are in small companies, that it is in the area of sub-contracting that most failures occur. At this end of the spectrum there is a high risk and volatility in becoming a sub-contractor. In the USA the failure figures for construction organizations show a similar pattern. The Census Bureau of the US

Department of Commerce showed in its 1987 census of the construction industry that the number of specialist contractors increased by 14% in the period from 1982 to 1987, rising from 299,433 firms to 341,484 firms. The number of firms going into liquidation was 3893 with $899 million of liabilities.

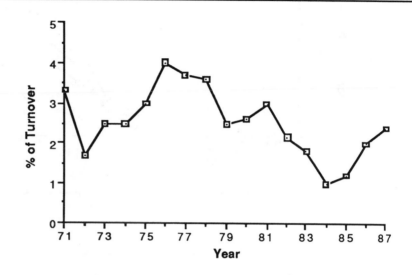

Figure 1.12 Profitability of Heating and Ventilation Contractors in terms of profits as a percentage of turnover.
Source: Heating and Ventilation Contractors' Association (reproduced by permission)

Organization of UK sub-contracting

The organization of sub-contracting into groups with a representative voice at the national decision making level has progressed since 1921, when the Federation of Associations of Specialist Sub-contractors (FASS) was formed. Since then other groups have been formed as the specialists have identified their various particular needs. There are now four groups, each under a particular 'umbrella' organization, three of whom have strong links with, and representation on, the bodies responsible for contract formation. As shown in figure 1.16, each umbrella organization's membership is composed of trade associations which represent smaller groupings around a particular specialism as well as individual member companies. In total there are about 200 trade associations with the 27 major associations being represented by the umbrella organizations. The umbrella organizations represent about 15000 companies out of the estimated total of 25000 sub-contracting and contracting companies in the UK construction industry.

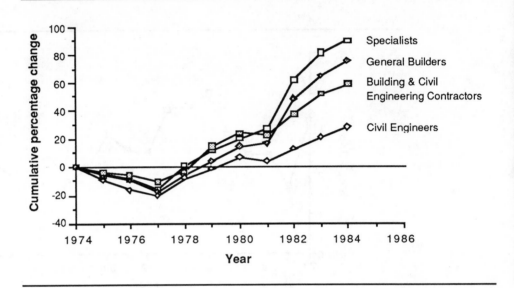

Figure 1.13   Changes in numbers of private contractors of all sizes in the UK, 1974-1984, Comparison by trade.
Source: Government Statistical Services. DoE 'Quarterly Bulletin of Housing and Construction Statistics'. and DoE 10 year housing and construction statistics'.

FASS (Federation of Associations of Specialist Sub-contractors)

This organization was formed in 1921 as a grouping of sub-contractors in order to afford mutual protection. The growth in influence of the organization was largely due to its director, Dennis Mallam, who retired in 1979. During Mallam's directorship up to 22 trade associations were represented and the organization had a strong voice in the industry. Since 1979 the organization has been split, first by the formation of CASEC (Confederation of Associations of Specialist Engineering Contractors) in 1961 and then by the formation of CASS (Confederation of Specialist Sub-contractors) in 1981.

The organization sees itself with two roles: speaking with an authoritative voice at national level on sub-contracting issues; and advising its members on specific issues and problems. FASS has representation on:

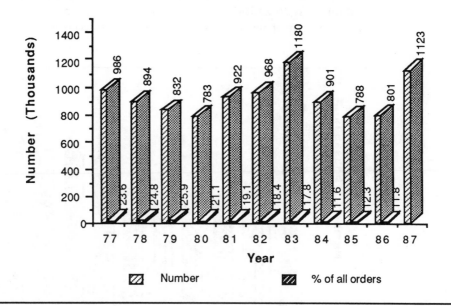

Figure 1.14   Construction Self-employed bankruptcies in England and Wales, 1977-1987.
Source: Department of Trade and Industry

❑ Joint Contracts Tribunal and various drafting committees and working groups

❑ CITB specialist building committee

❑ Joint taxation committee (with CASEC)

❑ Construction industry sector group - a civil service "think tank"

❑ PSA working groups - policy committee

❑ PSA consultative committee producing policy on sub-contracting (with CASEC and FBSC)

❑ Building industry working group

The stance adopted by FASS is to put forward the practitioner's view. This is done by choosing representatives who work at the practical level and who are dealing daily with the procedural and contractual issues. FASS are the only umbrella organization to have retained an MP as a consultant in order to maintain contact with legislation affecting sub-contractor's interests.

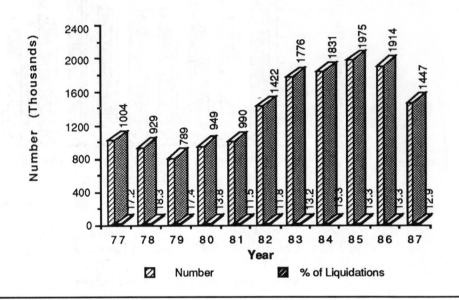

Figure 1.15   Construction companies liquidations in England and Wales, 1977-1987.
Source: Department of Trade and Industry

FBSC (Federation of Building Specialist Sub-contractors)

The FBSC is an umbrella organization within the Building Employers' Confederation (BEC), and its role is to represent the interests of sub-contractors from within the main contractors' camp. Its membership of the BEC ensures that it can meet to discuss, with the National Contractors Group (NCG), all of the issues affecting sub-contractors

FBSC has a dual role, firstly to act as an umbrella organization to provide representation at national level and, secondly, to act as the directorate for the trade associations in membership. The duality of membership means that companies are also members of BEC. The FBSC offers a wide range of services from specific advice to its members to representation on JCT committees. Members are, in general, looking for tangible services.

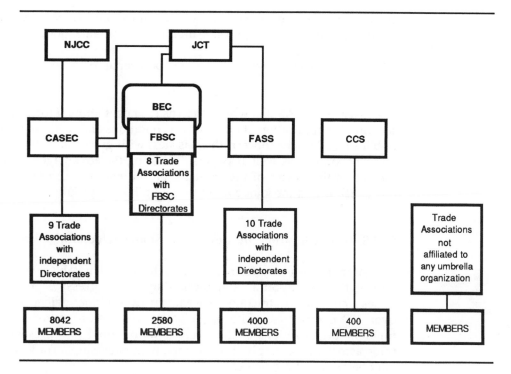

Figure 1.16 Relationship between sub-contractor governing bodies

CASEC (Confederation of Associations of Specialist Engineering Contractors)

This organization was formed in 1961 by the engineering members of FASS, who wanted a more specific representation of their interests. Centred around the trade associations of ECA (Electrical Contractors' Association) the HVCA (Heating and Ventilation Contractors' Association) and the SCA (Steelwork Contractors' Association), the organization now has a membership of 9 trade associations with a total of over 8,000 members. It is the largest umbrella organization, representing seven sub-contracting trades: heating, ventilation and air conditioning, electrical contracting, structural steelwork, plumbing, metal windows, lifts and suspended ceilings. Over its first 25 years CASEC has evolved a distinctive philosophy. The main points of that philosophy are:

❑ Sub-contractors should be organised strongly and separately from main contractors but, good relations with main contractors are being nonetheless vital;

❏   CASEC must be recognised by government and construction industry organizations as a responsible and reputed organization;

❏   CASEC must act responsibly and be prepared to compromise when necessary.

The majority of CASEC's work concerns the negotiation of forms of contract and securing the application of fair contract conditions and tendering procedures. Its publications largely concern these issues, with guides to the various standard forms of contract as well as papers discussing the advantages of nomination. It represents this sector of the industry on the national contract forums of JCT, NJCC, SBCC and the joint committee on building legislation.

CASS (Confederation of Specialist Sub-contractors)

This organization was formed in 1981 by John Huxtable. It is has a radical approach based upon a campaigning style of aggressive presentation of the issues facing sub-contracting. Since its formation it has gained a membership of over 400 companies. The organizational structure differs from the other umbrella organizations in that there is no pyramid of representation through trade associations, simply direct company membership. Trade associations can only be affiliated members.

Because of its relative youth and style of operation the organization has yet to achieve acceptance within the sub-contracting 'establishment'. It does not have any representation at national level on any significant organization's committees. However, through skillful use of the press, it has created a significant presence enhanced by its independence on many issues. By rethinking the needs of its members it is able to address a wide range of subjects in a way which gives those members a more direct and specific service. CASS produces a monthly newsletter which gives legal and contractual advice as well as analysis of onerous contract Clauses and general information about the trends in the industry.

CASS is working to promote its members as 'quality specialist contractors'. Quality and integrity are seen as desirable attributes to improve the performance of sub-contracting. These criteria are rigorously applied to its membership.

The main issues

There is not a coherent single voice representing the opinions of all sub-contractors. The only formal interconnection is a liaison group of CASEC, FBSC and FASS which meets three times a year. This is probably realistic given the wide range of interests and specialities. But the representation that exists, particularly on the JCT and NJCC, does not allow specialist contractors a presence on the contract drafting committees relative to their important place in the production process, and this may cause dissatisfaction with the current spate of contract difficulties. By splitting the representation, a strong power base, which truly represents the companies that actually do the work, has not been formed. There is a dissatisfaction voiced amongst the individual member companies that the larger umbrella organizations are not sufficiently independent nor representative.

> *"I would dearly love to see the three trade associations with which I deal all speak with a common voice as a common trade association"*
>
> Interview statement.

Summary

The many pressures of the last three decades, coupled with a reducing, but increasingly volatile workload, have forced a flexible form of working to emerge. This has encouraged the development of sub-contracting as has the increasing specialization of building technology and components which in turn has created a great diversity of manufacturing and supply organizations.

2

Sub-contracting experience in the 1980s

> A case study of a heating, ventilation, air conditioning, plumbing, and fire protection services sub-contract

General Overview

In every industry there are projects that go well and there are some that go badly. We can all learn by our mistakes, but so often we never have time to stand back and reflect upon why something went wrong. By the time the final account has been produced the staff are midway through the next project with its deadlines to be met.

This case study examines one work package on a project with the aim of seeing if there are any lessons that can be learned. It views a project from the perspective of the mechanical services specialist contractor.

The project in question has a final account value of approximately £7.9 million of which the heating and ventilating, air conditioning, plumbing and fire protection is worth £2 million. The project was built over a period of 30 months. For reasons of confidentiality, the dates and location of the scheme have been withheld.

The development had a chequered path. After detailed planning permission had been granted the development was sold to a pension fund by the developer. The building contract was between a development company established by the pension fund for the project and the main contractor.

The scheme was devised as a four storey speculative office building with a basement car park. The office building was divided into four separate but adjoining blocks. Each block was intended to be occupied by a separate tenant, but at an early stage of construction two of the blocks were sold to one owner who requested that the general contractor and the specialist contractors undertake the fitting out work as an extra to the contract, hence modifications to the structure and the engineering services were required. The new owner, a bank, wanted to install a network of computer terminals and a computer suite.

The tender was negotiated with a local contractor who submitted a lump sum fixed price bid on the basis of a JCT 80 building contract. There were nominated sub-contracts for the plumbing and the mechanical services. The principal relationships between the parties in the project are shown in figure 2.1.

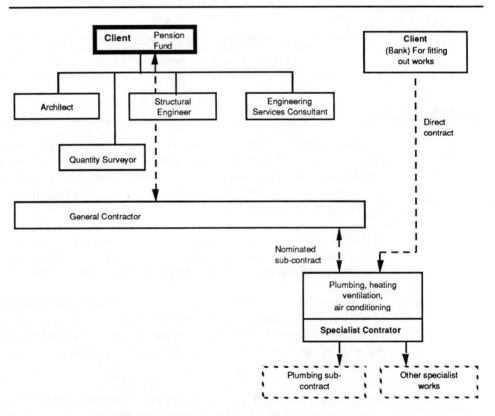

Figure 2.1   Schematic layout of relationships between project participants.

The specialist contractor

The company is a small to medium size heating, ventilation and air conditioning contractor which specialises in mechanical engineering services and has an annual turnover of £2.6 million. It has its own ductwork fabrication workshop and all the site installation is undertaken by the company's own craftsmen and operatives. Very specialised types of work, such as advanced energy management systems, would be installed by a specialist sub-contractor.

The company employs various craft skills and has a training scheme for apprentices. It is currently suffering from both a lack of available craft skills and the difficulty of getting craftsmen who are multi-skilled. Despite its size, the company has invested heavily in new technology and all the estimating and planning is carried out with the aid of desk top computer systems.

> A comment made by the company was "We want to be good engineers and provide a good quality service. We have become totally disenchanted with large project work because half the time has to be spent in interpreting contract conditions and responding to the mountains of paper generated by all the parties. It is unfortunate, but people seem to have become more interested in the paper and in reallocating risk rather than building the job."

The tender for the mechanical services

Tenders were sought on a JCT nominated sub-contract form. A number of drawings and a specification together with a list of free issue items were provided. The free issue items were intended to save the client money by the design team ordering the major equipment directly and then providing it to the mechanical engineering services contractor to fix. Four weeks were allowed for submission of tenders and an extension of four weeks was subsequently granted.

❑ A start date and completion date for each block was given. There was no information about the tenders for the electrical installation or suspended ceilings which are crucial trades because of the interface problems with the air conditioning.

❑ It was evident from the drawings that the design detail was incomplete. A clause in the documents to the tenderers stated that `the tenderer will be responsible for ensuring that the working drawings are fully adequate for the practicable installation of the works, and the *co-ordination of services.'* The difficulty was to price for this item when the tender drawings showed insufficient detail.

❏ A performance bond was requested from the specialist contractor but was subsequently not required after the client was informed of the cost of providing the bond.

❏ In the tender documents the specialist's offer was to remain open for acceptance until the 1st August. On the 28th August the services consultant wrote to say that they were instructing the contractor to enter into a nominated sub-contract agreement with the specialist contractor, but the start and completion dates for the contractor for the four buildings differed from those given in the tender documents.

	Tender		Acceptance letter	
	start	*finish*	*start*	*finish*
Block 1	9/9	31/12 (15 months)	2/9	1/3 (18 months)
Block 2	9/9	31/12 (15 months)	2/9	1/3 (18 months)
Block 3	15/7	1/10 (15 months)	2/9	31/12 (15 months)
Block 4	15/7	1/10 (15 months)	2/9	1/11 (13 months)

Table 2.1   Overall project timescale and its revision.

❏ No dates were given for the start and completion for the mechanical services contract within the project duration. The specialist was asked to confirm that the fixed price would not be affected by the changes in the contract dates (Table 2.1).

❏ The tender was accepted in the sum of £932,000 which included £105,000 of provisional sums to be expended as directed by the engineering services consultant. These figures excluded the cost of the client's free issue equipment which was worth approximately another £900,000. Whilst the free issue equipment was listed, there was no information about exactly what, where and how the equipment was to be fixed.

❏ The tender for the bank's fitting out work was accepted in the sum of £103,000. The work was to be completed in 22 weeks and was to be undertaken within the duration of the main contract.

The contract

❏ A JCT 80 nominated sub-contract was signed in September and an Employer/Sub-contractor agreement was also signed.

❏ The specialist accepted that in the certificates from the general contractor, the principle of "pay when paid" may be applied.

❏ In November, prior to the specialist commencing on site, the general contractor issued his *standard* general terms and conditions of sub-contract to the specialist - these were not refuted by the specialist. They included:

"We will not unload, check, store or place in position, neither will we take responsibility for, the safe custody, loss, damage, theft or any other such matters arising in connection with the sub-contractor's materials or plant. Should it be necessary for you to arrange for us to unload materials or plant we will do so without liability and all costs will be reimbursed to us."

"We allow you the free use of our scaffolding and hoists where already erected in connection with our own works together with the services of a regular hoist operator during normal working hours to your reasonable requirements. Such facilities would be available to cover operations within our building programme. The cost of retaining scaffolding and hoists beyond such periods due to delay on your part will be chargeable to you. You must satisfy yourselves at all times that this equipment complies with safety regulations and report any apparent breach of regulations to our Site Manager for rectification before use. Any special requirements for scaffolding must be specified in your tender and agreement on the provision therefore reached in writing between us, on or before the acceptance of your offer."

"We will not be responsible, prior to handing over completed works, to the employer under this main contract for damage to your work by persons, etc and require that you take all reasonable recognised precautions in the carrying out of your work to prevent damage to other existing work."

"You are to clear away all your rubbish and leave all work clean and perfect after completion."

❑   Whilst the above clauses might be fair and reasonable for a domestic sub-contract, they do conflict with certain clauses of the nominated sub-contract agreements which had already been signed by the specialist.

❑   Problems began to emerge on the contracts with suppliers of equipment who would not be bound by the terms and conditions of the main contract. In most instances goods will only be supplied based upon the manufacturer's terms and conditions of sale which can contain stringent clauses on payments and carriage.

Works on site

❑   Within one month of the specialist contractor starting work on site, problems were occurring due to the revised works for the bank. Figure 2.2 illustrates the complexity of the variations and changes that began to occur.

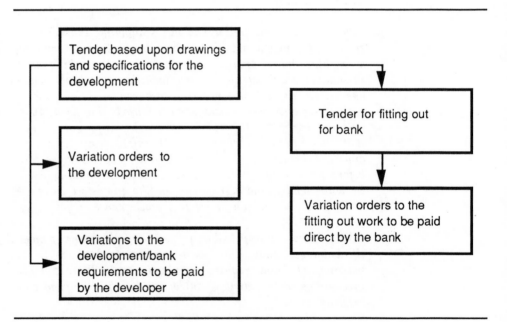

Figure 2.2   Details of variations and changes during construction

❑ Five months into the project the contract was behind time, caused by the late receipt of drawings and information and by the work in the basement having started in mid-winter.

❑ Over the next two months work on the specialist contract was being undertaken on a piecemeal basis and a lack of progress was reported as well as uneconomic work occurring. The client was becoming dissatisfied and the contractor was finding it difficult to make progress.

❑ The specialist's own sub-contractors, primarily the plumbing sub-contractor, were concerned about the lack of information interrupting the programme of work.

❑ The free issue equipment was beginning to arrive on site at about month six, some of it having been shipped from overseas; some items were received damaged.

❑ The work was rescheduled for the third time due to lack of progress.

❑ The revision of the drawings for certain sections reached revision z, having already exhausted the alphabet.

❑ In month seven the electrical specialist contractor complained about interface problems with the mechanical works and the lack of progress, and the mechanical specialist contractor complained about the delay of suspended ceilings work.

❑ The specialist contractor in month seven requested an extension of time in accordance with the contract conditions. No reply was received other than an acknowledgement of receipt.

❑ The correspondence between the parties became threatening and all the parties started blaming each other for the lack of progress.

❑ Nine months into the project and the contractor wrote to the specialist saying:

"Due to your apparent disregard to formalised work sequences, we find ourselves with little other choice than to warn you that any financial consequence arising from such total disregard will be deducted from any amounts of money which will become due in the future to your company."

"These costs will be calculated to ensure that all loss and/or expense suffered by this company and any other sub-contractors, as a direct or indirect result of your lack of progress in the areas made available to you at the above contract, will be recoverable."

"Your company has in the past given assurances that areas will be made available by these dates to allow access by following trades, but these dates have seldom been met. In particular, rainwater pipes are not being fixed to programme which has the knock on effect of delaying the drylining operations."

The legality of making such a deduction on the basis of 'set off' from the interim payments was not really considered by the specialist contractor.

❏   Twelve months into the contract and the specialist contractor again asked for an extension of time to his contract period. The request was acknowledged but nothing further was received. In order to expedite the works the specialist craftsmen were now working a six day, sixty hour week. The request for the additional resourcing had come from the contractor. The specialist wrote to the contractor saying:

"Our costs to accelerate the works to achieve completion, inclusive of overtime, have to be met. We are still hampered by continuing changes, subsequent late instructions, an incomplete fabric, and general site conditions which make progress difficult and expensive."

❏   The services consultant continued to produce design details as the work progressed on site. The distribution list for the drawings got longer as everybody became more concerned.

❏   In between dealing with the day to day technical queries the consultants began to receive letters seeking an extension of time and for the reimbursement of additional costs.

❏   The free issue items continued to arrive on site but were defective.

❏   Thirteen months into the contract the letters between the parties became very terse. Architect's Instruction number 456 was issued.

❏   The plumbing sub-contractor wrote to the specialist saying:

"Subsequently a second layout of Block 4 drainage, 1-50 scale, was issued showing substantial revision. ..... We are now informed

'unofficially' that this second layout is incorrect and that a third layout has been produced, but is not as yet issued. We feel that due to the above the only action we can take is to withdraw from the site until accurate and final working drawings are available."

The specialist was now becoming engaged in a dispute with his own major sub-contractor.

❏ The commissioning of the mechanical services was a much bigger task than first envisaged. Partial occupation of the building by the bank meant that commissioning had to be undertaken in phases; a two week exercise budgeted at £15,000 turned into a four week exercise costing £32,000.

❏ The co-ordination and interface problems between the electrical services specialist contractor, the suspended ceiling sub-contractor, and the specialist contractor got worse. Requests for information became continuous.

The outcome

❏ The main building contract was completed 46 weeks late and the client imposed payment of liquidated and ascertained damages. No extension of time to the contract had been granted to any of the parties.

❏ The bank fitting out work was programmed at 22 weeks and eventually took 30 weeks to complete.

❏ All the parties took up their 'claims position' with consultants being hired to assist in the preparation of claims for loss and expense. The general contractor imposed a 'set off' against interim payments certified to the specialist contractor. The specialist contractor prepared a claim for £172,000 against the contractor for uneconomic working.

❏ The bank became unhappy with the performance of the engineering services in their building and issued threats of legal action against the services consultant for possible failure in the design of the mechanical engineering services.

❏ Fifteen months after practical completion the final account for the mechanical services was settled in the sum of £1,170,000.

❏ The client imposed liquidated damages against the contractor and the contractor 'horse traded' a settlement with the specialist contractor, after it had been previously agreed to seek the appointment of an arbitrator. The settlement was for £63,000. Nobody felt happy at the outcome.

❏ During the progress of the work the specialist had received 20 revisions to the construction programme.

❏ The specialist's cost reconciliation after settlement of the contractual claims showed that the project covered its direct costs but made no contribution to head office overheads, and made no profit.

Lessons to be learnt and questions to be answered

❏ The project suffered badly due to an incomplete design prior to tenders being sought. The detailed design of the mechanical services was being developed as the work progressed which resulted in insufficient attention being given to planning the project. Could this have been avoided, given the nature of the specialist's own knowledge and design contribution?

❏ Detailed information is needed to describe any free issue equipment to be incorporated in a tender, a simple description of a chiller is totally insufficient for the specialist to price adequately. The specialist contractor should have sought clarification of the equipment at the tender stage. This raises the issue of package definition and the divisions of responsibility.

❏ The co-ordination and interface between the specialist trades contractors must be managed from the outset. Insufficient time was given to pre-planning the project. A start on site is usually wanted *yesterday*, but time invested in planning the whole project, both design and construction, would have shown up many of the deficiencies in the project's structure and organization. How many projects start in this way, with similar consequences?

❏ The modifications required by the bank were underestimated from the outset and an unrealistic promise was made to complete the work within the original construction duration.

❏ At the early stages of a project everybody is optimistic and assumes that the problems will be minor and easily overcome. The specialist does not study the fine print of the contracts, but the claims consultants do and it is only when events go badly wrong that the fine print in the contract is read and interpreted. It is not the fault of the standard forms of contract that the project went wrong, it is the fault of the people who were administering it.

❑ A £1,000,000 mechanical services project was a major responsibility for the specialist, yet the tender documents did not adequately reflect the risks being taken. Is it realistic to treat a major specialist contractor, whose work package represents 25% of the project value, in the same fashion as the labour only brickwork sub-contractor whose work may only represent 2% of the project value?

❑ The application of the right of 'set-off' was abused by the general contractor and there was very little, other than legal action, which the specialist could do to recover the amounts deducted from the interim certificate payment.

❑ The way that requests for an extension of time were repeatedly ignored by the design team showed a lack of concern for the contractor and all the other specialists and sub-contractors affected. Is the design team the most appropriate group to assess the consequences of failure?

These problems and issues were specific to this situation but they raise many questions which sub-contractors face each day. The underlying reasons for this failure and the failure of similar projects are many and complex. The main reason is that few people have a broad view of the construction process and the part everyone must play in order to ensure that the failings described here are not the norm for the industry.

3

Sub-contracting and the process of construction

As sub-contracting has been developing so has the way in which it is used in a project. It is a complex development and needs to be explained carefully if many of the problems now faced by sub-contractors are to be understood. This section develops a simple model of the construction process in order to explain the role of the sub-contractor in the process of constructing a building. It will be used throughout the report to illustrate the role and responsibilities of the sub-contractor in the UK and other countries, as well as the effect of the various contractual relationships that can exist.

The 'manufacturing' process now occurs off-site and much of the knowledge of the process is held by the specialist manufacturers and the technical specialists within these companies. It is this fact, more than any other, which has had such a revolutionary effect upon the way in which the construction process must be managed. For a construction project to be created all of the required knowledge must be brought together to consider the problem. It is this fact which has yet to be fully appreciated by everyone involved in the industry.

Figure 3.1 illustrates the first part of the basic process model of a building project. This process model underlies all modern building. It is also common to all the construction industries of the developed world where buildings are primarily assemblies of components. In practice, the model is extremely complex in that the stages in the process overlap and fuse together and the process for each component runs in parallel with the others, but usually at a different rate. Therefore, to understand the full implications of the model, analysis must take place at two levels - the project level and the individual component level. This discussion considers the component level in the context of the project needs.

Site assembly

The modern process of building is largely that of assembling predetermined components in predetermined locations. Each location is specified by the design. The order in which the components are assembled is dictated by the support and fixing system. Each component must be securely fixed to a support as it is expected to remain in position for anything between one year and several centuries.

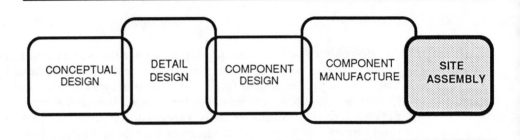

Figure 3.1   The site assembly task

Construction is thus a sequential operation of assembling components which provide the support for subsequent components. This method of building, which has not materially changed for centuries, has a restraining influence on the way buildings are constructed.

Whilst in principle this applies to all buildings, the detail of every building is different, which of course adds to the rich visual scene of our towns and urban landscape. This does, however, lead to immense problems, the first of which is that the skills and experience learnt on one building cannot be transferred directly to the next building. This is a minor problem where the design draws upon the traditional or well established technologies and craft skills. It is a major problem when the building project is using new skills drawn from other industries or where components are created for the particular building.

The second major problem is that the design must state every situation which is unique in order to provide clear instructions as to the precise details of the component assembly. This dependence upon instruction from the design has grown in recent years through a combination of factors. As component based assembly has increased, so the flexibility to modify and adapt the components during the installation process has reduced, thus necessitating the identification of unique situations for the design and manufacture of the components. Also, because the components and junctions between them are specific to the project and the technology is non-traditional, any modification or adaptation using site skills may void the design liability. This has onerous implications for the long term building liability. In an attempt to avoid this liability the contractors and specialist assemblers do not undertake any design activity or site modification, which could be construed as design, that has not been approved by the design team.

Up to this point the site assembly process has been considered as an entity. In practice, it is a process involving many organizations, each of which is contributing to a particular aspect of the process. The modern building site is therefore a collection of different organizations, each with their own task and company philosophy brought together in a combination which is probably unique to that situation. Each is attempting to maximize their opportunity in respect of the company's objectives. The fact that many of the specialist organizations are contributing design expertise, which is heavily modified through interaction with the design team, means that every specialist sub-contracting organization is serving three masters with different, often conflicting, objectives: the design team; the site assembly production process; and their own company objectives.

The growth in the use of manufactured components and the technology available to the manufacturer has had an effect upon the way in which the more traditional materials such as brick and concrete are being used. The site assembly process uses a wide range of materials and components, each of which has its limitations in terms of the precision with which it can be made. If these limitations are not understood, enormous difficulties will arise when attempting to join the components together. This is compounded when components manufactured to industrial tolerances, ie +/-1mm, are to be joined to components manufactured on site with much wider ranges of tolerance, ie +/-12mm.

Components manufacture

With the increasing use of manufactured components the site assembly process has become very reliant upon the availability of the component in that it should arrive on time, be technically correct and of the required quality. How the component is manufactured is, therefore, an integral part of the total construction process model. The first thing to recognise is that the UK component industry does not manufacture a consistent product which is then put into a stock to be drawn upon in any quantity by the site. Except for a few products such as cement and Fletton bricks, whose demand is fairly constant and can be produced continuously, virtually every other component is made in relatively small batches. This has an extremely disruptive effect upon the manufacturing facility as it cannot gear up for highly efficient, single product runs. Each component requires a slightly different machine setting or sequence of manufacture in order to achieve the particular specification.

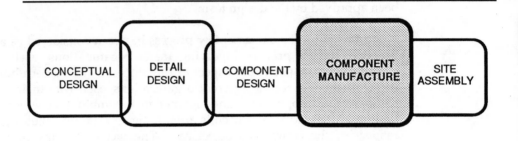

Figure 3.2   Component manufacture by specialist sub-contractors

In order to resolve the economics of small batch production the component manufacturers have adopted several techniques, the most significant being extended time and sub-contracting. Extended time enables the work flow through the manufacturer's shop floor to be scheduled, as economically as possible, by having large time buffers between each stage of the manufacturing process. This allows the maximum utilization of the machinery before it is changed for the next batch. It also allows the workforce to be scheduled as economically as possible by giving them a constant flow of work by the creation of backlogs of work at each stage. Sub-contracting enables the manufacturer to smooth the peaks of demand by passing the peak workload to another manufacturer who will also be a small batch manufacturer. Alternatively the sub-contractor may be a specialist in his own field, in which case the whole component design and manufacture process will be implemented.

In all events the time for the whole manufacturing process must be estimated correctly. In order to make this accurate estimation the manufacturer must consider the type of manufacturing process, the likely use of sub-contracting and their manufacturing process, and the assembly of the sub-assemblies into the final component.

An example of this process was the sequence of cutting and polishing granite cladding for a prestigious building in the City of London. The stone was hewn in sufficient quantity for the complete building in a quarry in South Africa. The quarry was only open for one month in every six months. Each block of stone was taken by train to Durban for shipment to Italy, where it was taken to specialist stone cutters in Travertina, north of Rome, for cutting and polishing. Once each individual piece of granite was prepared it was packed into a crate holding approximately 1.5 tonnes, loaded into containers and then brought overland to the channel ports and eventually delivered to London after another sea crossing and overland journey. The whole sequence took 52 weeks from placing the order to the delivery of the first piece of granite.

The significant lessons to be learnt from this example are that many products and components are manufactured outside the UK. This is not surprising given the UK's mercantile and trading history. Extensive transportation is necessary for the majority of components and the method of delivery tends to favour the economics of transport rather than the eventual site handling. Hence 40 tonne containers arrived on site containing 1.5 tonne crates of stone which could not initially be handled efficiently.

The increase in the use of manufactured components has exacerbated the problem of managing the construction process. Many specialist contractors and specialist sub-contractors are involved in manufacturing the components. The nature of the demand however, necessitates that production is in small batches all over the world in order to take advantage of manufacturing skills and economies of scale. Time is required to smooth the work flows in order to minimise the disruption and time is taken to move the components between the different geographic locations as they are prepared for final delivery. Failure to appreciate the complexity of this process or the potential for disruption will mean that the components will not arrive to suit the construction process and so the site activity will be disrupted.

Component design

The growth in the use of manufactured components has largely been a result of the development of the materials and manufacturing technologies now available. Each development in technology increases the specialism and specific capabilities of the manufacturer. In many cases the specialization has been turned to specific marketing advantage in that the manufacturer, whilst presenting a product range, will emphasize his ability to adapt the product in order to supply specific components to the particular building. This implies that the manufacturer has an extensive knowledge of his own capabilities which is unique to him. The designer of the building must also discuss and evolve the specific adaptation of the component in a continuous dialogue. In practice, in order to achieve market share and sales of the product, the adaptation of the product becomes so extensive that the final component is very remote from the original concept of the product.

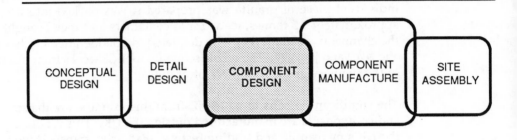

Figure 3.3   Component design by specialist sub-contractors

Designers are now acutely aware of their limited knowledge of the details of the product technologies and are turning more and more to the use of the specialist manufacturer's knowledge. There are two stages to the development of the knowledge for a specific project. Each stage involves the development of detailed drawings. The first stage is the development of a detailed interpretation of the designer's intentions. In many cases the designer will indicate the scope of the problem and seek the specialist contractor's advice on the way in which his products or expertise can be used to provide a solution. The discussion will be extensive and many drawings will be produced to evaluate alternative solutions to the problem. These detailed component drawings will then be used by the designer to develop the detail of the whole building design in the sure knowledge that the technology is the best that is available. Once the principles have been agreed the second stage will proceed with the development of the shop drawings. These are the drawings that the manufacturer will use to manufacture the individual pieces and components. It will depend upon the product and manufacturing process as to how extensive this stage will be. Structural steelwork, for example, will require that every connection is calculated and the bolt and weld locations be specifically detailed.

The process of detailed design described above, whilst relatively straightforward for a single component, becomes considerably more complex with each interaction between components. It is not until the detailed drawings or even the shop drawings are produced that the exact nature of the interaction can be identified. Particular problem areas arise when one component provides the fixing base or even the fixings for another component. The base component cannot go into manufacture until the details of the fixing requirements are available from the other manufacturer. Considerable information interchange is essential between the specialists, their sub-contractors and other specialists in order to achieve the resolution of these detailed issues before manufacture can commence. This process of interchange is further complicated by the need to obtain the design team's approval as the details are being prepared prior to issue for manufacture.

Detail design

It is a characteristic of the majority of recent UK buildings that their quality comes from the attention to detail taken by their designers. It is therefore essential that the design team is extensively involved in every aspect of the detailed design process. Design practice in the UK is probably unique in this respect, certainly when compared with practice in the USA and some parts of Europe. As buildings have become more complex the specialist's involvement in the detailing has become greater to produce the information required by the site process. To achieve this level of sophistication the demands of the designer must be communicated to everyone who requires the information or is affected by it, which implies an extensive transfer of information between them.

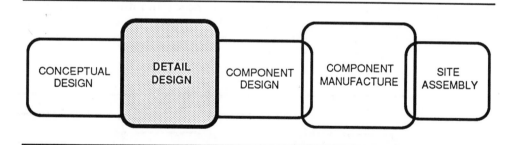

Figure 3.4   Detail design by specialist sub-contractors

If pre-made or off-site manufactured assemblies are employed in the construction, the problem is exacerbated. These components cannot be readily altered on site, therefore, every non-repetitive situation must be identified and detailed. Further difficulties arise where the components are made to factory standards and tolerances, as frequently the supporting structure is required to meet the same standards of accuracy and fit. If it does not, then the final standards of finish are unlikely to be achieved. Consequently, the detailing of the support structures to the finishes increases in complexity such that every detail is considered in depth in order to maintain the consistency of finished appearance. The creation of the detailed design is thus one of continuous development with input to the decision process from members of the design team. This input reflects the implications upon the support systems and the aesthetics with input from the specialist contractors as to their ability to resolve the design problems given their particular products.

Conceptual design

During the conceptual design process certain aspects may be explored in great detail to determine the technical viability such that even final working drawings may be fixed. The implication of this is that when either the full economic or practical analysis is complete fundamental changes may be required which are impractical because the design has advanced too quickly and is in a too committed state. The design process, therefore, requires vast resources of information, together with stringent evaluation. In many instances the specialist trade contractors are involved at this stage to give their advice to the design team. This is done in good faith and not paid for, but the level of detail that is often demanded requires a significant investment in design and management time. Many specialists are questioning this investment as there is no direct return in new contracts because the majority of these are subject to competitive bidding. There is, therefore, a reluctance to contribute the information that the design team needs and it is not until the sub-contractor is contracted that a dialogue will commence. However it is often too late at this stage to achieve the smooth incorporation of the data into the project and so delays and chaos occur.

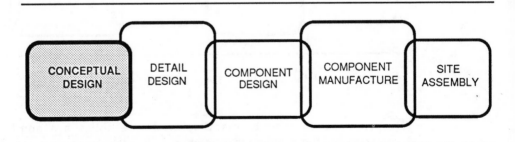

Figure 3.5   The complete construction process

James Lansdell of architects Kneale and Russell, has found that product selection and contractual procedures are two recurring problems on many of their cladding projects.

The product selection problem arises on projects where the need for good appearance and quality is married to a limited budget. A statement often made by clients in this situation is: "we don't want it to look like an advanced industrial unit in its most basic form but we don't suppose we can afford a highly sophisticated system like system X?" A less well developed system has to be adopted and developed .....

.... A last ditch tour of Scotland was undertaken to find a tried and tested system of a comparative price to the brickwork but which also had the aesthetic appeal we were seeking. The tour was a success and a system was located which had been used by a new town Development Corporation. However, there were still problems to overcome: the integration of doors, windows and roof covering and the potential cost increased as a result of these factors.

This is the stage when the contractual problem arose. Most systems are only supplied by the manufacturer or agent. Although part of the system the ultimate 'design' is the responsibility of the erector, who at this point is, like the the supplier, not yet appointed.

The main contract was to be the result of a competitive tender and the client did not wish to nominate the sub-contractors or suppliers. It was, however, permissible to name the product in the tender document but the erector of the system had to be selected by competitive tender. This made it impossible to develop the detailed design of the system prior to a main contractor being appointed. This meant there could be a potential delay to the start of the contract, and accurate costs would not be known until tenders were received.

Fortunately all external apertures and the roofing element formed part of the one element for tender purposes. All of the component parts of this element - the external envelope - could be 'named' and their erection carried out by one domestic sub-contractor who would be part of the main contractor's competitive tender.

The development of the external envelope design virtually ceased during the design process when input from the erection sub-contractor was required. Very little additional information was forthcoming from the supplier at this stage since, without financial commitment, he was unwilling to give further information on his product. When tenders were received and a contractor appointed the responsibility for the design of the system passed to the main contractor.

Passing the Buck   Building 21/4/89

Summary

The process described above is so complex, so bound in traditional practices and not understood as a whole by many people, that it is very difficult to co-ordinate it to achieve a single objective. Each stage in the process must be used for what it is best able to achieve. A project is a collection of a myriad of specialists called together to solve a particular problem. This, whilst being an inordinate strength in terms of innovation, craftsmanship and art, is also its weakness. Diversity into small units of expertise does not allow an integrated design and manufacturing process based on high volume. The location of the final assembly depends entirely upon the site location, thereby destroying the continuity of the manufacturing process every time the building is complete, as the team must physically move to another piece of land and start the whole process again. Experience is accumulated by the expert or specialist somewhat fortuitously depending upon circumstance and luck. Very few people actually see a project through from start to finish, thus it is very difficult to develop a strategic view of the whole process.

It is hardly surprising that the industry fails in the eyes of the public and its clients. There are so many points at which it can go wrong. Each specialist makes the assumption that his work is perfect and that the people who have executed their work before him know what they were doing. Inevitably there is variation in performance. A system which is composed of small segments, relying on perfection at each stage, is doomed to failure, unless its weaknesses are understood and controlled.

4

The implications of various contractual options on construction

The strengths and weaknesses of the existing 'standard' contracting methods must be analysed to understand the implications of the various contractual approaches on the sub-contractors. This is not a clause by clause analysis but a review of how the contract binds the parties into a framework of responsibilities and whether those responsibilities are realistic given the structure and needs of the building process. There are very few reliable statistics about the types of contracts in used in the industry. The Junior Organization of the RICS has conducted three surveys which give an indication of the changing emphasis in the types of contracting practice now being used (see figure 4.1). The survey showed that over 50% of the projects were undertaken on lump sum contracts using bills of quantities. Design and build is showing a consistent increase, whilst the use of management fee contracting, based on the number of projects it is used on, has declined since 1987, although by project value it has increased.

Five interconnected points are relevant to this debate:

1   The client wants the building on time, within budget and in accordance with the specification.

2   The contractor and consultants want to make a profit and enhance their own reputations in the building process .

3   The nature of the building process is that 100% success in 1 and 2 above cannot be guaranteed.

4   Arrangements have to be made which identify where the risks for failure should be carried.

5   The growth of sub-contracting has had a significant impact on the balance of the risks between the contracting parties.

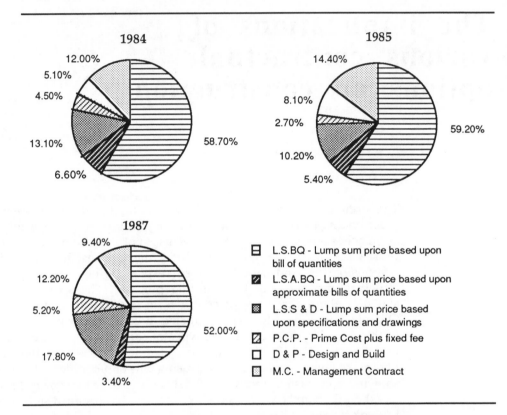

Figure 4.1   Trends in type of contract forms. Percentage of contracts by value. (1984-87)
Source: The third of the JO(QS) surveys - contracts in use in 1987

In the attempt to attain these criteria, the determination of clear responsibilities for ensuring that the actual processes to achieve the finished building are often overlooked. Furthermore, the industry does not help itself as there are considerable gaps between theory and practice as shown in the box below.

Theory	Practice
The design is complete prior to tenders being sought where lump sum contracts are used	The design is rarely complete often due to client pressure for an early start
The contractor will build the project using some specialist contractors	In some cases up to 95% of the work is undertaken by specialist contractors
In selective tendering none of the bidders knows the identity of the other tenderers	It is not difficult establish who is on the tender list
Bids sought from sub-contractors at the tender stage will not be subject to a Dutch-auction	Some Dutch-auctioning does occur with sub-contractor bids
The contractor will seek permission to use sub-contractors	Some permission is sought, but seeking permission in every instance is rarely practical
All sub-contractors are assumed to be of equal importance, the painting sub-contractor is not treated differently to the structural steelwork sub-contractor	The larger sub-contractors are sophisticated in their management systems as the contractors
The client and architect can make changes to the scope and extent of the work during the construction phase	Any changes to the work once it is in progress affects the construction programme and causes inconvenience
An extension of time to the contract will be dealt within a prescibed manner	No extension of time is straightforward and agreement can be long and painful
The final account will be settled and agreed in a reasonable time	Final accounts take a long time to agree and settle
All the parties will behave in a professional and trustworthy manner	Everybody does, until something goes wrong and then it is always somebody else's fault

In the following analysis of contractual systems the process diagram developed in chapter 3 has been used to illustrate how the various responsibilities are distributed. The diagrams are intended to indicate the responsibilities embodied in the principles of the contract. It is recognised that each contract is tailored to the specifics of the particular project and the lines of responsibility are drawn accordingly. The position of the lines in the diagrams is indicative of the boundaries. Where they are drawn as overlapping there is no intention to indicate proportions of work or a specific proportion of the project time scale, but to indicate a shared responsibility of perhaps varying scale when comparisons are made between one contract system and another. Finally, it is hoped that this approach, rather than the conventional heirarchial diagrams, enables a clearer appreciation of the relationship between responsibilities and the task that is trying to be managed, a link that is often overlooked.

The lump sum bid form of contract

This type of contract, such as JCT 80, is created on the premise that the design team completes the design prior to the bidding of the project. The implication is that a clear responsibility can be established for the production of the design and that the design will be as near to 100% complete as to enable a specification and bill of quantities to be produced. The bill of quantities allows tenders to be produced on an equal basis from which to select the contractor. Another assumption is that the contractor can work unimpeded and has total control over the work and is totally responsible for the provision of the materials and workmanship to the quality specified by the architect.

As a concept this may have been practical when a building could be designed by architects and engineers and executed by the tradesmen and operatives directly employed by the contractor. But, as figure 4.2 shows, this view does not fit the modern construction process. The increased use of the specialist contractor has meant that the simple separation of design and construction is not possible.

We are not suggesting that all specialist contractors have a design input. Obviously plastering, painting and brickwork contractors are not involved in producing design details, whereas curtain walling, electrical and mechanical services, structural steelwork and cladding frequently play a major role in the design process. Specialist contracting in this context therefore refers to those with a design input.

If the responsibilities of the design team are shown in relation to the whole process they cover the brief, conceptual design and working drawings. It is the partial development of the working drawings by the specialist contractor, which are used by the design team to develop the design, that breaches the separation between the designers and the builders.

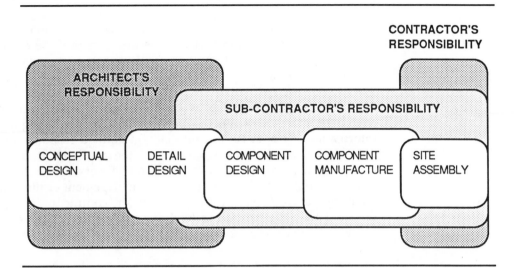

Figure 4.2 Specialist contractor's design, manufacture and assembly responsibilities under a lump sum bid contract.

In practice, the specialist contracts with the main contractor who, under the traditional arrangements, is responsible for the provision of materials and workmanship. The main contractor is not responsible for the provision of design information. In fact, most contractors take great pains to exclude any reference to design in main and sub-contracts. The dilemma is that to complete the design the specialist contractor must contribute to its production, but there is no contractual chain of responsibility to facilitate this. The consequences of this are:

❏ The specialist contractor's loyalty is divided, because he must satisfy the design input he is requested to achieve by the design team but, in so doing, he is taking a design risk. The main contractor is primarily interested in site performance, therefore, any delays in providing the design information are between the specialist contractor and the designer. The designer has no direct call on the specialist sub-contractor and it is, therefore, only through goodwill and sensible practice that the design is delivered. In terms of the construction process this is a very weak system and can lead to confusion over the responsibility for its management.

❑ Contractually the specialist's design information only becomes available when the specialist is contracted with the main contractor. This system often does not recognise that the timing of the design process, which works in a different sequence to the construction process.

❑ Because the specialist contractor is offering a manufacturing capability, the discussion between his designer and the design team emphasises his skill in interpreting the design requirements and in fact contributing to the aesthetic design.

❑ The three way division between the design team, specialist and contractor effectively splits the responsibility for producing the design information. It does so, not into clearly defined responsibilities, but into a confused situation which leaves the specialist at risk and the design team with no clear capability for delivery of the design. The chain of command is split and thus the management of the delivery of the design is weakened. If the parties in the contract do not recognise this split then the contract is likely to fail because of the complexity of interaction necessary to complete the design of a modern building.

In conclusion, this form of contract does not adequately recognise that specialist contractors input extensively into the design process and that the design team's work cannot be completed without their input. A division of responsibilities exists, which has no meeting point. In fact there is a no-man's land where the specialist contractor and consultants are at considerable risk. A number of attempts have been made to overcome the conceptual difficulties of the traditional system, for example, by modifying the main contract provisions to require the general contractor to take responsibility for part of the design, or by introducing a warranty to be provided by the sub-contractor direct to the employer in relation to design. The reduction in nomination has been another attempt to put the management responsibility for the production of the specialist's design into the main contractor's realm but, for this to work, the timing must allow the design interactions to occur and issues of design liability must be resolved.

The management fee form of contract

The dissatisfaction with the lump sum bid has caused the industry to consider alternative ways of procuring contracts. One of the most significant developments, certainly in terms of growth and market share, has been the 'management contract'. There are many variations upon the principal theme. The method's historical development is somewhat clouded but may perhaps be ascribed to one source, Arup Associates, the integrated multi-discipline design practice of Ove Arup and Partners. They sought to obtain the involvement of the contractor within their concept of the integrated team. It is said that after a visit to the USA they evolved the idea of adapting the US construction management ideas within the UK context. At that time conventional contracting was going through a radical change in employment practices because it was becoming the management of sub-contractors. It was no great jump in logic to extend the concept of nomination to all sections of the work and confirm the contractor in his role of managing the sub-contractors.

The great strength of management contracting has been the ability to move the contractor alongside the design team, who can then benefit from his knowledge of the scheduling and operation of the site assembly process. Designers had also recognised that the sub-contractors were, in many cases, developing significant expertise which they would like to use. The management contractor provided a convenient vehicle to employ the specialist and obtain his input at a point to suit the designers.

The responsibilities, as shown in figure 4.3, are largely unchanged from those under a lump sum contract except that by appointing the management contractor earlier the interface between designer and contractor moves much closer together. The management contractor will not accept any design responsibility so the input by the specialist contractors must still be warranted directly to the client. This means that the design team is still responsible for the delivery of the design to the contractor with the same consequences for failure as under a lump sum contract.

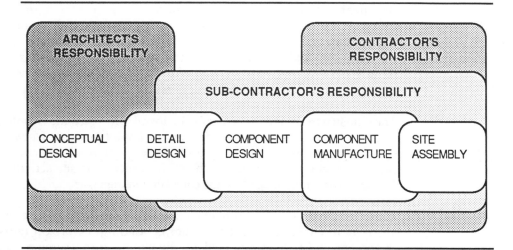

Figure 4.3   Sub-contractor's responsibilities under a management contract

Whilst there are many strengths in being able to bring the contractor into the project at the earliest opportunity, the fundamental issue of a split and the resultant confusion of responsibility for the complete process still exists. It is for this reason that amongst those clients who have used management contracting there is a growing awareness of its weaknesses and thus increasing disenchantment with its use.

Professional construction management

In the United Kingdom construction management is understood to mean the integrated management of design and construction. This is not to be confused with the way the term is used in the USA where it means the management of the total process. However, both systems have recognised the complexity of the design and construction interface and have attempted to evolve systems to manage it.

Professional construction management is offered as a service and bought by clients in the same way that design or quantity surveying are bought. A professional service for a fee. The professional agreement is fundamentally different to a commercial agreement. The construction manager is not taking the same commercial risks as a contractor and the financial rewards will reflect this. The client is taking the maximum risk with this form of contracting because he has no form of guarantee on construction price or duration when he embarks on construction. The client obtains a management skilled in the construction process without any of the conventional contractual constraints. In practice the contractor's skills are absorbed into the project team and become the client's directors and managers of the work, as shown in figure 4.4. The important point is that the construction manager does not assume the conventional contractor's risks of default by the package contractors. Unlike the lump sum and management contracts, where the architect is responsible for issuing instructions and interim certificates, here the client issues them with the advice and guidance of his designers and managers.

The construction manager's role is to ensure that the whole design and construction process is managed as an integrated process. He will assist in the definition and procurement of each package of work although the package contractor will be contracted to the client direct. The site activities will also be co-ordinated and managed by the construction manager.

Although the process is a form of contracting it would be more correct to call it a situation in which the client or owner is building the project himself and thus the responsibility is totally held by the client. It is, therefore, a construction system for extremely competent and knowledgeable clients who understand and are prepared to take the risks involved. The client has control of the design and the work and can spread the risk through the contractual agreements.

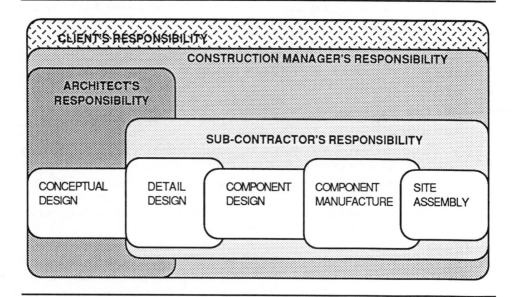

Figure 4.4   Sub-contractor responsibilities under a construction management contract

The sub-contractors and professionals who work under this contracting system must also be quite specialised because they must have an extensive knowledge of construction and the design process in order to be able to maximise the potential of contributing their knowledge at the inception of the project, alongside the design team. Where this has occurred it has placed huge demands on the sub-contractors. They are being asked to challenge all of their existing practices and become totally committed to the project. They are fully exposed to the demands of the project as there is no main contractor to undertake the detailed planning and organization. Each sub-contractor has to effectively become a main contractor. In some cases conventional main contractors are becoming package or trade contractors.

A worrying feature that emerged from the survey of specialist contractors was that some of the specialists were unaware of the fundamental differences between management fee contracts and professional construction management. One comment was "It is more or less the same contract under a different name". In terms of responsibilities and liabilities it is not the same.

The British Property Federation (BPF) System

In practice, construction managers require the risk to be taken by others, but there are few clients who are skilled enough to do it successfully. However, there are many clients who are frustrated by the alternatives and, through the British Property Federation, they have voiced their corporate view that they could devise an alternative, viable procurement system. In 1983 they produced a manual which described their proposals. This was followed in 1984 by a form of contract produced in conjunction with the Association of Consultant Architects (ACA).

In principle, the system attempts to resolve the responsibilities for the design and construction processes by defining their scope. The intention is that the architect defines 80% of the design before seeking a single lump sum bid for the whole construction. Any remaining design is completed by the contractor. The pedigree of this system is clearly in the conventional US contracting system where design and construction is divided and divorced. This system has become less fashionable in the USA with the growth of fast track construction, ie overlapping design and construction.

As shown in figure 4.5 the responsibilities are clearly drawn around the stage of detail design. The BPF system manual is ambivalent as to quite where the limits of each party's responsibilities are drawn. In practice it will depend upon the involvement that the design team requires of the specialist contractors in the detail design process, but where the specialists' design input is significant the BPF system may be inappropriate. There is no mechanism in the BPF system for this involvement outside of the contractor's contractual relationship with the specialist.

The system attempts to emulate the US system whereby the contractor bids for the work on the basis of a specification (performance specification) and set of drawings. He then has total responsibility for delivering the building. In the USA this system is used to design down to a price and maximise the design/ production/quality trade off. The contractor is thus in control of the information and product supply, which can be scheduled in accordance with the site assembly requirements. Under one alternative of the ACA form of contract the contractor is required to provide any necessary amplification of the information which is supplied. If he does provide drawings he is to warrant that they comply with the performance specification and that the part of the design carried out by him is fit for the purpose. The design is also to be submitted to the client's representative to obtain his approval. In practice, this leaves the contractor open to a design liability, unless he seeks an appropriate limitation clause, against the background of an overall design to which he did not contribute. If the client's representative is the architect then the contractor also has to provide a design which complies with the architect's finished details, details that may not have been stated.

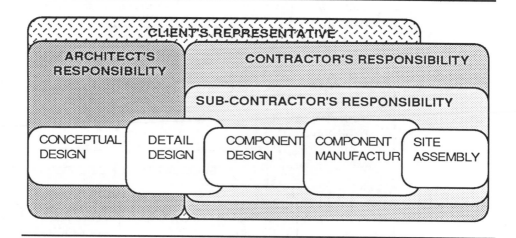

Figure 4.5   Sub-contract responsibilities under the BPF form of contract

Unless the parties enter into the spirit of the BPF system then it could get bogged down in a very acrimonious and detailed argument with each party trying to limit its liability.

Design manage and construct system

This form of contracting system is a relatively new innovation. It has been developed by a few contractors in an attempt to enable them to provide a completely integrated package to their clients. One popular way has been the design and build or package deal, whereby the contractor offers a complete price for designing and constructing the building to meet the client's needs. For complex buildings the amount of design required to define the project has usually outweighed the time and cost advantages of the package deal.

A variation on the package deal theme is where clients are seeking to use their own independent consultants, but with the advantages of the whole process being managed as an entity. There are two methods for achieving this advantage. The first is that the contractor employs the design consultants as if they were specialist contractors under a management fee form of contract. The second method is where the contractor takes the design being developed by the design consultants and prepares a lump sum bid including the consultant's fees. If he is successful in this bid he will embrace the design team as if they were nominated sub-contractors. Both methods have the effect of giving the client a single point interface to the process through the contractor who in turn is responsible for delivering the complete design and construction package, as shown in figure 4.6.

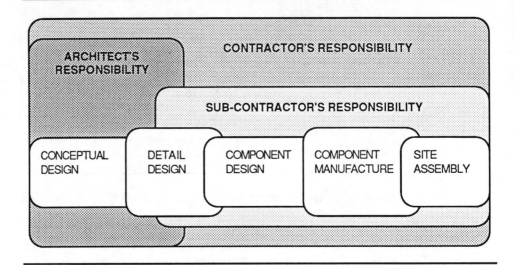

Figure 4.6   Sub-contract responsibilities under a design, manage and construct form of contract

This form of contract is attractive to designers who have the burden of the management function removed from them and so that they can concentrate on designing the building. However, it also means that they have to submit to the discipline of the construction based programme and they no longer have the last say on matters affecting design. The contractor is in the driving seat and delivers the project against a price. He will try to maximise the design by having access to its development and can introduce modifications to achieve high site productivity. The system can lead to acrimony with the designers refusing to accept responsibility for a design if their own wishes have been overruled by the contractor. The ultimate sufferer would then be the client.

The alternative method of management (AMM)

This method of designing and managing buildings was pioneered by Ray Moxley of Moxley, Jenner and Partners in response to the desire of the architect to be in direct control of the whole process. In practice the architect replaces the contractor and becomes his own construction manager. He designs the building, appoints the trades and supervises the work on the site, as shown in figure 4.7. It is a method reminiscent of the great days of Wren who designed and managed the building of his churches and cathedrals in London.

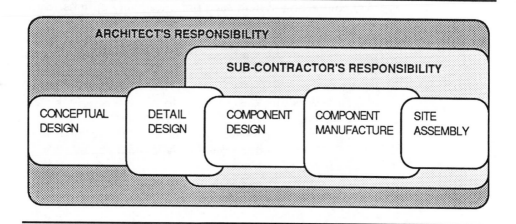

Figure 4.7   Sub-contract responsibilities under the AMM form of contract

Most of the projects that have been undertaken using AMM have been small in scale. They have also required an architect who is committed to the concept. In return the experience has been enormously satisfying because the architect is involved in every aspect of turning his vision into reality. It is particularly the direct communication with the tradesmen that is so attractive. Design details can be discussed and the results will take account of the tradesman's knowledge of his craft and the materials being used. The designer and the trades, therefore, build a close mutual regard for their respective skills.

The trade packages are contracted direct to the client, therefore, as with construction management, the client takes a substantial risk with this type of arrangement. Design and construction are, however, combined under the single responsibility of the designer who is able to maximise on the construction knowledge of the specialists. As long as the project is contained within a good, effective system of cost planning and control the inevitable enthusiasm of the architect can work to the client's advantage. Whilst there are considerable personal satisfactions in working in this way, it is considerably more demanding for the architect and it is for this reason that only small projects have been undertaken using AMM. Interestingly the systems used in France and Germany reflect a greater involvement of the design team in managing the building process along the lines of AMM.

Summary

There is no perfect contracting system. Whilst these are the basic, generic, types of contracting system each provides many options. Systems which have not been covered specifically are: GC works series I and II, which are the main government forms of contract; JCT with contractor design, which is another form of contract giving the contractor control and responsibility for aspects of the detail design; the ICE civil engineering forms; and FIDIC for international contracts. All of these 'standard' systems are open to detail adjustment by clients, surveyors and designers, with the lawyers looking on to assess the clauses covering liabilities. It is this flexibility which is the danger. Whilst there are a few optional clauses in the standard forms it is not intended that that such forms should be subject to wholesale changes.

A formal contract exists as a record and confirmation of the duties and responsibilities between the contracting parties. Contracts are often regarded as an end in themselves and are being adapted and experimented with as if they could change the steps which are implicit within the sequence of the design and construction process. They can only affect the relationships, often adversely, between the parties executing the steps in the process. If the contract is used to force one party into an untenable situation then it is unfair and the threatened party cannot be expected to perform as expected. This may seem obvious, but the clauses in many building contracts are used to lever a particular response from the contracting parties. The sub-contractor has very little chance of negotiating the conditions and articles of agreement and sometimes he appears to fail to perform in accordance with the letter of the contract. In many instances he has no choice because the contract, and those who drew it up, have failed to recognise the structure of and responsibilities required within the project.

When choosing and preparing contracts the client must ensure that he is protected and that his objectives are achieved. He must clearly understand how much risk he is prepared to take and actively manage his risk. The following are the main problems to be overcome in order to ensure that whatever procurement system and form of contract are chosen, the importance of the facts of the integrated nature of design and construction is recognised:

❑ The approach must provide a proper and effective means of controlling cost and ensuring that the project is completed on time.

❑ The approach must ensure that there is a continuous management responsibility for producing the design information in accordance with the requirements of the site assembly process.

❑ The approach must recognise the contribution to the development of the detail design by the specialist contractors and the architect's need for that contribution in order to complete the design.

❑ The specialist contractor has important knowledge about the best ways to use his materials and components, that knowledge should be harnessed as early as possible in the design team's thinking.

❑ The system must recognise that the civil law liabilities in the event of failure of the design of buildings are, at present, extremely onerous and that contractors and suppliers will only enter into agreements which recognise this in limiting or mitigating their liability.

❑ The contract must be drawn up in such a way that it accurately reflects the client's intentions and objectives. For example, the modified basic JCT standard form of building contract has been used in a construction management context. The principles of the two approaches are incompatible. The practice of taking a standard form of contract as the starting point for extensive manipulation, without careful thought, puts everyone at risk.

❑ Each clause in the contract must be compatible with the objective of ensuring effective management of the whole process. Contracts often contain contradictory clauses.

To produce a building effectively requires the close co-operation of designers, specialist contractors and contractors all working to the same objectives. Unless the management framework, through their contractual responsibilities, recognises the contribution that each has to make in the complex, interconnected design and construction process, then the project will falter.

Are there any lessons to be learnt from the philosophy of building contracts used overseas? The Japanese General Conditions of Building Contract, which are widely used, states in the general principles (Article 1), '*the owner and contractor shall perform this contract sincerely, through; cooperation, good faith and equality.*' Perhaps this is a clause worth considering for the UK market! The Chinese contract shows a similar influence of eastern culture; it states, '*all items not found in this contract will be deliberated upon in a spirit of mutual understanding and trust.*' Many of the difficulties in the UK arise precisely because of the lack of trust endemic in our adversarial system.

5

Contract conditions

The JCT has developed a family of standard forms of contract which attempts to cover most eventualities on construction projects. The forms are produced by concensus and the task of getting all the parties to agree is a daunting one. The JCT forms have attempted to reflect the changes that have occured in construction. The JCT family of contracts have kept pace with the changes in contracting practice but, clients are very diverse and they have very different needs which result in amendments being made to the standard forms. Sometimes the amendments impose conditions that appear onerous to one of the parties. No party has to accept an onerous condition of contract, but the commercial pressures will often outweigh the judgement and the contracts are signed in "the hope that the worst will never happen". In the interviews we conducted it was apparent that many of the smaller sub-contractors signed contracts with onerous conditions because if they did not, somebody else would. The commercial environment is competitive and the effort to maintaining workload results in more risk being taken.

A viewpoint on onerous contractual conditions

Developing experience over the years has led consultants and contractors to be confident in amending the standard forms of contract. FASS President, Phillip Simpson, at a conference in London (Building, 1989) listed 24 major contractors who had made up to 15 amendments to their benefit. He was highlighting a growing concern amongst the specialist contracting industry that possible abuse of the standard forms of contract could potentially lead to damage to the long term interests of the Industry and its clients. The formalisation of this worry was produced by CASS (Huxtable, 1988) but due to the diffracted nature of the specialist's umbrella organization's response, the contractors have the upper hand. We must point out that the contractors often have onerous conditions imposed upon them by the client and they then transfer some or all of the risk on to the specialist contractor.

Some of the most significant issues arise in the following areas:

❏ The sub-contractor is presumed to know the detailed conditions of the main contract and all its requirements and provisions without actually seeing the full document. In most cases the contractors extract those sections which they deem to be relevant for the pricing and tendering task and cover all other items under a general "catch-all" clause. In theory there is an opportunity to inspect all the documents at the offices of the contractor. In reality the pressure of estimating and pricing often leaves insufficient time to inspect the documents.

❏ The sub-contractor is responsible for co-ordinating his design with that of others. Whilst architects are responsible for co-ordinating the total design, it is becoming more common to pass some of this responsibility into the hands of the specialists who are preparing working and shop drawings, particularly in complex areas with multiple specialist input. Difficulties arise when the sub-contractors are unaware of the extent of the interface which occurs when all the sub-contractors involved in the detail are not available because they were not appointed at the right time.

❏ With the changing role of the contractor the sub-contractor is being increasingly required to co-ordinate his site operations with those of the other sub-contractors. In practice this requires that the sub-contractor employs a higher grade of site management.

❏ The most significant clauses in domestic sub-contracts concern the provision for payment. The most common changes are the so called 'pay when paid' clauses. This means that some of the cash flow risk is being shared with the sub-contractors. In essence the main contractor will not pay the sub-contractor's invoice until he has been paid by the client for the interim valuation in which the sub-contractor's work was included. This means that payment can be made anywhere between 5 and 11 weeks after the work is done, if the respective accounting practices mesh together. As most normal business terms are payment within 30 days of submission of an invoice, then the sub-contractor may be supporting a negative cash flow. The most difficult aspect is the lack of certainty with regard to payment, because the sub-contractor does not know when the contractor has been paid, together with the additional risk that either the client or contractor might fail, both of which will prevent payments to the sub-contractor.

❏ Another area of payment practice which is open to dubious practice is the right to set off, against the interim payments to the contractor, any claims for failing to meet the full conditions of the sub-contract. The standard forms of contract give the main contractor a right to deduct a justifiable claim, but there is also provision for the sub-contractor to challenge this and if necessary go to an adjudicator to quickly resolve the matter. However, in some instances the arrangements for the adjudication are deleted and much abuse is practised through the inherent leverage of threats of a claim .

Under a new standard set of conditions of tender, members of the 60-strong Federation of Brickwork Contractors will increase prices by 15% for any pay when paid clause. The brickwork contractors say they will charge an extra 5% for any departure from their standard conditions of tender. They will charge a further 10% for any clauses on set-offs or cross claims which have not been agreed and delays for payment will be subject to interest at the rate of 15%. An extra charge will also be made for performance bonds.

Contract   Journal   27/4/89

❏ It is increasingly common, particularly on management and construction management types of contract, for the sub-contractor to be responsible for the protection and making good of his work, no matter when it was installed and whoever causes any damage. These clauses are intended to protect the main contractor who may be unable, through his minimal staffing levels, to oversee everything. In practice the early trades find this unworkable and it is open to abuse by the later trades, which often leads to set off claims. Whilst it is designed to ensure that the British workman takes some degree of care over the work of others it is generally impractical and a 'bad rule'.

Summary

Whilst the standard forms of contract are designed to ensure that the sub-contractor completes the work satisfactorily, the range of amendments and additional conditions ensure that the Sword of Damocles is perennially over his head threatening retribution through cash and legal penalties. It is not a way of working that is conducive to developing a harmonious working relationship.

Most sub-contractors, when faced with a heavily amended contract, either return it, or heavily qualify their bid or return the bid based upon their own standard terms and conditions. There are other situations where it is prudent to ignore the contract conditions altogether, for instance:

> *"We are involved in a plethora of contracts. The contracts that we have worked with at .... are absolutely horrendous and I'd never send them to be evaluated by our legal people because if I did and took the contract after their advice they would have me committed to an institution."*
>
> Interview statement

The general attitude that prevails is that if the company is well organized and knows what it is doing then its exposure and risk are minimized. Where a relationship based on professionalism and respect for each others contribution is established, then the words in the contract are rendered superfluous.

Roger Wakefield of Nabarro Nathanson suggests that (letters, Building 10/2/1989):

> *".... it would be a mistake to consider that the conditions imposed on sub-contractors in construction are harsher or more onerous than in other industries. In any case, the treatment of sub-contractors by other sub-contractors demonstrates considerable hypocrisy with regard to complaints about the onerous nature of provisions imposed by contractors on sub-contractors."*

Which provoked a response from H R Howard, President of FBSC, in reply to Mr Wakefield (letters, Building 10/3/1989):

> *"The main contractors, through their delegates endorsed the findings of the good practice panel of the NJCC which was subsequently published - 'that it is beneficial to the industry as a whole for standard forms to be used' (when placing sub-contracts). Therefore I fail to see that it is necessary for contractors to add their own onerous conditions. .... it only results in increased rates, the final cost of which has to be met by the client."*

6

Warranties and the specialist contractor's liability for design

> *"A small but significant number of professional practices are actively encouraging the use of design warranties and guarantees which impose liabilities on sub-contractors far in excess of those agreed as reasonable by the Joint Contracts Tribunal"*
>
> Richard Tully,
> Chairman
> National Joint Consultative Committee for Building
> (reported in The Chartered Builder)

Contracts and warranties

An area of increasing concern for all the parties on a building project is the liability for latent defects. Disputes about liability for defects in buildings have occured for centuries and they will continue to do so. For example, when Sir Christopher Wren was building the wing on Hampton Court there was pressure to finish the project quickly. Such was the pressure and the skimping that, at a crucial stage of construction in 1689, a considerable section of wall and roof fell down, causing the architect to be hauled before a board of inquiry. Like any modern day counterpart, Wren argued that there was nothing wrong with the design, but that his builders had been using sub-standard mortar while he was off the site attending to another contract, a cathedral at Ludgate Hill His pleadings were accepted and he got away with it.

The disputes that occur today are very similar, one of the biggest difficulties being to ascertain who was responsible for any defect that occurs. Simply described, the contractual arrangements between the client and those involved with a project are shown in figure 6.1 and listed in Table 6.1.

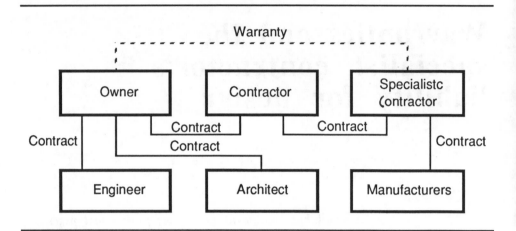

Figure 6.1   The contract and warranty arrangements

While the contractual arrangements are relatively straightforward, under English law the advantages of a clearly defined contractual liability are offset by the requirements of privity of contract which confine the benefits and liabilities to the immediate contracting parties. As a result, an owner may not claim under the law of contract against a sub-contractor with whom he is not in contract, but this does not affect the position in tort where there might be tortious liability.

The difference between a contractual and a tortious obligation is that the first is imposed by an agreement between the parties, whereas the second is an imposition of reasonable responsibility by the law itself. The real basis of the tort problem is the measure of damages and the area of doubt regarding any recovery in tort of economic loss, ie, cost of repairs.

Interestingly, in the United States of America, remedies have been provided for the consumer by dismissing the requirement for a contractual link between the consumer and the producer. The Consumer Protection Act 1987 represents a small step down a similar path in the UK, but the building industry is only marginally affected by the legislation.

Parties	Contracts
Owner in contract with the design team (architect, engineer, quantity surveyor)	Conditions of engagement issued by the respective professional institutions
Owner in contract with the contractor	JCT standard form of contract or other standard form
Contractor in contract with specialist contractor	JCT standard form of sub-contract, or DOM/1 and DOM/2 standard forms, or contractor's own conditions of contract
Specialist contractor in contract with manufacturers	Manufacturer's conditions of sale and the sub-contractor's conditions of supply
Employer in contract with the specialist contractor	JCT employer/sub-contractor forms of warranty

Table 6.1   The parties and the relevant contracts

The precise scope and existence of duties and liabilities under the law of tort is extremely uncertain and presently undergoing change as a result of recent cases. To overcome some of the ambiguity, various warranty forms have been introduced into the building industry which provide a contractual link between parties who would not otherwise be in contractual relations.

The decision of the House of Lords in D and F Estates Ltd. and Others v Church Commissioners for England and Others (1988)(2A11 ER 992) gives cause for a major rethink on the question of liability of those involved in the construction of new buildings. Also it affects those who will suffer loss if buildings are defective. The case concerned the construction of a block of flats, completed in 1965. The main thrust of the case concerned defective plastering work that was undertaken by a sub-contractor. The response to the D and F Estate's case is that the person who acquires a building is likely to ask for more warranties.

The employer/sub-contractor agreements provide a warranty for the client when the specialist contractor undertakes some aspect of the design as part of the contract works, for example:

> *'the sub-contractor warrants to the employer that he has exercised and will exercise all reasonable skill and care in the design of the sub-contract works in so far as they have been designed by the sub-contractor, so as to ensure that such designs are safe and suitable for their purposes.'*

A contractual relationship exists, therefore, between the employer and the sub-contractor. Many of the individual warranties now being produced by employers are not limited to design. They require a full warranty for the performance of the works undertaken by the specialist.

The aim of the warranty form is laudable, but in the building industry there is an assumption that nearly all of the design work will be undertaken by the architect, structural engineer and consultant environmental services engineer. As building components and equipment have become less standardised and often more complicated, so the reality is quite different. It is the specialist contractor who has the detailed knowledge of how the different types of equipment would fit together into the building and it is the specialists who provide the shop and installation drawings for approval by the architect.

Having signed a form of warranty there is the question of what constitutes design. In addition to the obvious design work, tasks such as selecting particular materials or components for specific functions can also amount to design. The tender documents may contain, in addition to the word design, such expressions as: 'development of design', 'performance specification', 'selection of materials' or 'fitness for purpose', each of which will imply a partial responsibility for design.

An architect or an engineer who has agreed to design a building may not, in the absence of express provision to the contrary in the forms of engagement, avoid liability for negligent design on the ground that the design was delegated to another person. Under the RIBA Conditions of Engagement, the terms of the architect's appointment may permit such delegation, but generally this does not apply to the delegation of the design details often provided by specialist contractors. In this situation, theory and practical reality part company. The specialist contractor, such as may be used for the supply and fixing of suspended ceilings, has detailed knowledge of all the different types of detail. He will work to the architect's layout and detailed design requirements, as well as co-ordinating with the electrical and mechanical services contractors. Inevitably, certain of the shop drawings will be produced by the suspended ceiling specialist contractor and approved by the architect. However, none of the architect's conditions of engagement refer to the role of shop drawings. A further difficulty arises in the amount of cover provided by the specialist contractors' professional indemnity insurance for design liability; assuming the specialist contractor had such a policy. A professional indemnity policy will normally cover against negligent design but will not cover a contractual undertaking regarding suitability of the design. Therefore, the value of such warranties has to be questioned, when they are not backed by the professional indemnity insurance.

The examples brought to the attention of the NJCC's Good Practice Panel had certain notable characteristics; in particular the documents were clearly devised to bring to the employer all the benefits of nomination whilst avoiding the responsibility which the procedure necessarily placed on the client and his professional team.

"Great effort and considerable expense has clearly been devoted to the drafting of the documents which are often complex and incomprehensible to the layman unversed in the law". Consequently, such warranties and guarantees carry grave dangers for the unsuspecting, and a sub-contractor who fails to consider the ramifications or insurance implications before tendering may be making a first step towards commercial oblivion.

The Chartered Builder.   October 1988

The increase in demand for warranties

Collateral warranties are frequently the products of legal advice to owners and developers and tenants of various types of commercial property. The aim of the document is to bind a party in contract to another party with whom he would not otherwise have any contractual relationship. The warranty provides a means of pursuing a legal remedy where one might not exist in its absence.

There are many cases where the client/developer is little more than an agent for a funding agency who will eventually become the owner. In such a case the contractual responsibility of the architect to the developer is of little value to the eventual owner if latent defects arise.

Collateral warranties are becoming commonplace, but the legal position is confused. The warranty forms require a full warranty for the performance of the contract and the design to be given to the employer, and any third party to which the warranty is assigned. This can include those funding the project, future tenants and a whole plethora of people who may not even be known at the time the building is being constructed. This is an extremely onerous requirement for any company to give. However, from the point of view of any future tenants they are becoming increasingly reluctant to accept responsibility for latent defects under full repairing leases. Their view is that, "we did not build the building, so why should we carry the can!"

Funding agencies and freeholders look for warranties from every conceivable source and owners are requesting warranties from most of the specialist contractors on projects. The interests of each of these different parties vary and, in order to provide a realistic level of protection, the terms and conditions included in the warranties will vary.

The agreement will inevitably be prepared in the form of a deed, which means it will be executed under seal and the limitation period for any claim is twelve years rather than six years.

It is beyond the scope of this report to discuss the merits and demerits of collateral warranties. Suffice to say that the specialist contractors are currently being required to sign warranties that are going beyond the bounds of common sense.

Onerous conditions and responsibility for design

The standard form of employer/sub-contractor agreement relating to any works where there is a design component is clear and unambiguous. The difficulties occur when conditions are changed and onerous conditions are added. For example, it is common to replace the word 'reasonable' with 'utmost' and to include words to the effect that "the sub-contract works will be entirely fit for their intended purpose." It would not be possible to obtain professional indemnity insurance cover for such open-ended obligations. Furthermore, insurance companies are very sensitive about companies who enter into contracts where more than minor or insubstantial amendments are made to approved forms, without having notified the insurers.

The sub-contractor is also required to sign a sub-contractor's deed of collateral warranty in favour of the employer, which, in part, may state:

"the sub-contractor warrants to the employer that he has exercised and will exercise all reasonable skill and care in the execution of the sub-contract works including:

a) The design of the sub-contract works in so far as the sub-contract works have been or will be designed by the sub-contractor so as to ensure such design is safe and suitable for its purposes, and;

b)   the selection of materials and goods for the sub-contract works in so far as such materials and goods have been, or will be, selected by the sub-contractor; and the satisfaction of any performance specification or requirement is included as part of the sub-contract works."

Another example of an onerous condition would be:

"In so far as the sub-contract includes any design, the sub-contractor undertakes to use the reasonable skill and care associated with his expertise in the preparation and execution of such design and to ensure that such design will be safe and suitable for its purpose and shall indemnify, and keep indemnified, the contractor against any loss or damage suffered, howsoever caused, arising out of, or in connection with, any defect in or the unfitness of such design and against all claims or demands by the employer or any other person in respect of such loss or damage."

The specialist contractor and the contractor are both in the position at the tender stage of not knowing exactly what parts of the design will be undertaken with the aid of shop drawings prepared by the specialist. Most importantly, the specialist must indemnify, and keep indemnified, presumably in the form of insurance, both the contractor and the owner against loss suffered as a result of failure. Few of the smaller specialists carry such insurance today, and few will carry insurance for claims that might occur in five years time as a result of a project failing in some way.

Other onerous conditions that sometimes appear concern the period of cover, where the insurance is required to continue for a period of 15 years rather than the 12 years under the deed - obviously taking the liability period to the maximum under the Latent Damage Act 1986. The insurance requirement imposed on contractors and sub-contractors also stipulates that the insurance should be related to a maximum per claim, with no aggregate limit on claims in any particular year. This type of cover is virtually impossible to obtain in the current insurance market. The normal is a maximum per claim with an aggregate limit on the total of claims in any year.

The lines of responsibility can become complicated as shown in figure 6.2. If we now assume that a particular design detail has failed and a latent defect has occurred, the owner has two main concerns: to get the defect put right; and to ensure that someone else pays the costs. The owner must establish:

❑   why and how it failed;

❏   whether the failure was a result of bad design, bad workmanship, poor specification, or material failure;

❏   the extent of the remedial works required to remedy the defect;

❏   who was responsible for the failure;

❏   the extent of the period of limitation;

❏   and the loss suffered as a result of the defect.

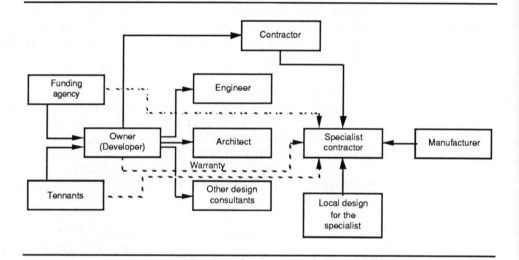

Figure 6.2   Contract and warranty arrangements

Whilst all the parties will want to co-operate with the owner to ensure that the defect is remedied, the practicalities of the situation are that the owner will be confronted with all the parties blaming each other. In the first instance, all the parties will pass any claim to their insurers, assuming they have arranged cover, who will generally, as a matter of course, refute liability. The situation can be further complicated if a developer has sold the completed project to a new owner who was not involved in any aspect of the building work. The new owner will not, therefore, have a contract with the design team. Against this background, it is hardly surprising that the legal principles surrounding the subject do not exhibit the same degree of consistency and coherence that one would expect of the law on such an important subject.

The limits of liability are not clear and many specialist contractors are confused and ignorant of the true position. They are unclear about how long they have continuing liability after completion of a project. The warranties owed to a third party provide adequate cover as long as the third party remains solvent. Inevitably, when a major problem occurs, the smaller companies cease trading and the client is still left with defects which no party is prepared to rectify.

The limits of liability of the professions and contractors in the construction industry and the extent of both obligations and liabilities clearly need careful examination.

Shop drawings and installation drawings

Specialist contractors have recently become more involved in providing shop drawing information. For instance, the lift manufacturer will provide and install the lifts in accordance with the consulting engineer's and architect's drawings, the performance specification, the catalogue information and the installation drawings which will be approved by the consultant services engineer prior to installation work taking place. The lift manufacturer will ensure that his standard lifts are customised to fit into any building and meet the performance requirements. In this case the lift manufacturer will supply installation drawings which show the finished product. At the opposite end of the spectrum the groundwork specialist contractor may have devised a particular method of working that would avoid using sheet piling in part of the site, thus saving time and money for the owner. The shop drawings would show the temporary works and the revised method of groundwork support and would need to be approved by the consulting structural engineer.

Under the JCT standard forms of contract the status of shop drawings and installation drawings is not clearly defined. The architect/engineer may approve a shop drawing for conformance with the design principle; however, if the detail fails as a result of bad design, the client may be able to sue the person who approved the drawing if the approval was given negligently.

Approvals are becoming a major issue in the UK because the liability issues are unclear. In the USA the status of shop drawings is stated in the various standard forms of contract. The American Institute of Architects form of contract A201 states:

> *The Architect will review and approve or take other appropriate action upon a contractor's submittals such as shop drawings, product data and samples, but only for conformance with the design concept of*

the work and with the information given in the contract documents.
Such action shall be taken with reasonable promptness so as to cause no
delay. The Architect's approval of a specific item shall not indicate
approval of an assembly of which the item is a component."

The problems have occurred when the architect has approved the sub-contractor's shop drawings for a particular design detail that has subsequently failed. The client has joined the architect with the contractor and the sub-contractor in the action to recover damages. The question of liability is still unclear and, despite the clarification in the contract, disputes continue to arise and the American Sub-Contractors Association currently have a working party attempting to clarify the liabilities of the various parties.

Similarly, in the UK, concern has arisen because of the few cases where a latent defect has occurred in the building after completion. Often the client has no choice other than 'joining' all the parties in the action to obtain redress, because of the uncertainty as to the cause of the defect and the overlapping or uncertain legal responsibility.

The terminology for drawings is not adequately defined in the building industry. A shop drawing is generally a drawing which is issued by a contractor or specialist contractor and shows the detailed assembly process that can be used in a workshop and on site. Although it has been issued by the contractor or specialist, the drawing may have been prepared by an independent consultant acting for those parties. However, some specialist trades use shop drawings for fabrication purposes only and prepare installation drawings for assembly on site. To avoid further confusion, it is recommended that the types of drawing and terminology should be clearly defined.

Figure 6.3 expands the design and indemnity aspects of the contract and warranty arrangements shown in figure 6.2. As well as signing a standard form of sub-contract, which will contain a clause outlining the responsibility for design, it has become common for the contractor to seek a warranty from a domestic sub-contractor on behalf of the client.

Looking to the future

There are short term solutions to the difficulties surrounding warranties. Contractors and sub-contractors need to be diligent when scrutinizing the warranties they are being asked to provide. The requirement to provide more warranties is likely to increase and the position is likely to become more complex. All parties should seek expert advice and train staff to fully understand the risks the company is carrying.

Design and Liability	Indemnity
Architect produces detail design and performance specification for a specialised work package.	Architect is covered by professional indemnity insurance against loss as a result of negligence.
Specialist contractor develops the design details for the work package and submits the details for approval by the architect. If no in-house design capability is available the specialist will sub-contract the design to a local designer.	Specialist contractor will be covered by an 'all risks' insurance policy that will not cover loss as a result of defective design undertaken by the specialist contractor. Separate policy is required for design liability insurance, which is expensive, and is not frequently held by specialists. If a local designer is used, the specialist will not necessarily ensure that the designer has professional indemnity insurance to cover design liability.
Manufacturer recommends standard design equipment to the specialist for incorporation into the shop drawings, stating it will meet the requirements of the performance specification and can be incorporated into the project.	Manufacturer will have product liability insurance.

Figure 6.3   Design activity and associated indemnity needs

A small step has been taken recently towards a revolution in the building insurance market in respect of latent defects. Consideration is being given, by some insurance companies, to offering single project insurance in respect of latent defects. The scheme is based upon the developer insuring the building structure for 10 years after completion, whereby he can recover the cost of damages for any defect without going to court. The insurance company would employ consultants to check and monitor the drawings and construction. However their likelihood is that whilst the latent defect would be repaired by the insurance company, the search would continue, by the insurance company, for anybody that might have a liability and that could include the specialist contractor.

7

Production and productivity

It is a fact that the British construction worker is neither indolent nor stupid. Comparative studies with the construction industries in other countries have shown him to be less productive at his task but have also shown that his task is complex and not well managed. Studies have shown that the average construction worker is only productive for 40% of his time, the rest is spent moving from one task to another or waiting for materials and instruction. In other words there would be no shortage of skilled manpower now or in the future if the tasks were well organized and managed. Whilst this has been known since the middle of the 1960s it is only with the advent of the 'managed' form of contract that it has been considered seriously. Value engineering and buildability focus on three stages in order to raise productivity: the design of the task to be performed; the working conditions for the operative; and the organization of the working environment. The problem is largely overcome by management action in the early design stages. This has led to an increasing use of complete work packages so that the sub-contractor can manipulate the design of his work in order to achieve the high levels of productivity that are needed. It can be done. There are numerous examples which show that a radical rethinking of construction activity assembly, eg lifts, has streamlined the site activity.

Task design

The average time spent per visit by a tradesman to a traditionally built house is one and a half hours. On a recent major office project the dry lining fixers visited the same core over twenty times in order to complete the cladding whilst working around the other trades. This is a typical example of task design that is the outcome of conventional design team thinking. The cladding operations were subsequently rationalized down to six operational visits with no loss of design performance once the sub-contractor was given the opportunity to propose alternative designs. Task rationalization can only occur during design if it is to have any significant effect. The sub-contractor must be in a position to contribute to this type of thinking, both by having the right expertise available and being employed at the right time. His own designers must be fully aware of the practical problems faced by the site teams. In addition to satisfying the design teams' requirements they must achieve the following to enable the site teams to work steadily and continuously:

❏ Rationalize the work into large units in order to increase the work time from a few hours to a one day minimum of continuous activity for each operative.

❏ Remove tasks which are unnecessary by redesign. Every manufacturer is now involved in this task in order to increase his productivity. The Japanese are masters of this.

❏ Remove the need for a dependency on other contractor's components to provide the fixing base, thus removing the need for breaks and revisits.

❏ Assemble, off-site, complex areas which are difficult to organize on-site, eg, toilet pods, lift cars, lift motor rooms, plant rooms, bathrooms and air handling areas.

Task design should ensure that on each single visit that the work can be finished in its entirety. No pride can be taken in a piece of work that is never finished. It is usually the result of poor organization that pieces are left off or the item is considered to be so minor that it is left 'until someone happens to be going past that site'.

By consistently applying these rules the motivation of the workforce will increase because they can see that their task has been considered carefully by a 'management who cares'.

The working conditions

Site work is wet, dirty and physically demanding. Yet we expect work of a high degree of accuracy to be created in these conditions. To an extent it would be difficult to change many of these intrinsic problems, but if the workforce is to be retained, then the site conditions have to be comparable with other industrial conditions. Once again there are examples where this problem has been taken seriously. At the Broadgate development in the City of London, for example, the whole of the area around the buildings was concreted and cleaned daily. Even during winter it was possible to walk comfortably around the site in town shoes. The problem has two aspects: work area control and task completion control.

Work area control

Most sites look chaotic, with materials and men spread around like confetti leaving very little scope for a gang to work coherently. As the organization is increasingly placed in the hands of the sub-contractors, they must develop a sensible strategy to maximize their productivity. The biggest single problem is interference from other trades trying to work in the same area, or crossing the area to get to their own work. Control must be imposed around the gang trying to work and the rules are quite simple:

❏   Physical demarcation of the work area is required to impose the discipline. Barriers or tapes will provide an elementary discipline for other trades trying to enter the area.

❏   A single sub-contractor must have control of the work area for the duration of his work in the area and the duration of the work in any area should be such that a smooth flow of gangs through each area is achieved.

The discipline described above is hard to achieve and requires thought from designers when detailing the work. As shown by the example below it can be done with surprising results.

In order for this system to work a vigorous site management discipline is required, which may or may not be applied by the contractor. But, increasingly, as the performance is being sought it will become more widespread and the sub-contractors will come to demand it. It is not a one way street, everyone must be committed. Again the rules are simple:

❏   The preceding trades must have fully completed their work;

❏   The work area must be clean and tidy;

❏   The preceding work must be accurate;

❏   All materials and components must be available;

❏   All plant and equipment must be available;

❏   The work area must be checked as safe and all safety provisions and equipment must be available;

❏   All access routes must be clear of the area and other trades must not be permitted to cross the area.

Fitting out Phase 2 of the Broadgate development for Shearson Lehman American Express has not been an easy task. Alan Crane, divisional director for Bovis Construction is adamant that the 16 month contract worth £54m to fit out 31,600 m² is a record breaker. "........ In our view, our progress has been better than anywhere else in the world. And that goes for the best the Americans have ever done." The secret is that a series of interrelated practical ideas has been adopted.

The work is cordoned off into three distinct zones that are tackled sequentially in the four areas of the floors. First is the "ceiling" work - installing the air conditioning, sprinklers, lighting, security and suspended ceilings. Only once the ceiling is completed can work start on the raised floor and below deck services. Last, there is the area in between - erecting the partitions and adding the furniture and fittings. A strict design rule was laid down: on no account were the services to stray from one zone to another. Fixing light switches to partitioning is taboo. Instead automatic sensors are used, or the switches clustered in as few areas as possible, generally around the service cores. One of the fundamental things the designers and trade contractors had to take on board was the rigid division between the three zones.

Another factor in developing techniques to streamline the fitting out programme was the full size mock up built before work started. This was invaluable in helping to perfect the three zone method of working. "The workforce embraced the concept at an earlier date than their employers."

That'll do Nicely, Building 12/6/87

This imposes a tight management discipline on the the sub-contractor's management to ensure he is organised. In order to achieve these requirements the sub-contractor has to make sure that the whole of the off site delivery system is working perfectly. Material and equipment supplies must be ready at the work point so that the operative does not waste valuable time looking for the components. The example below shows that this can be achieved, resulting in this case in an increase in productivity of 40% per man.

> The materials management system is clear and simple. It hinges on dividing the floor into four colour coded areas. All the materials have both floor number and colour code marked on the delivery note by the supplier. Materials are only called off by the trade contractor when required. All movement of materials is done at night by the "graveyard shift". The sub-contractors contribute to this gang, whose job it is to transport the next day's supplies to the correct workplace on the floor. The same goes for removing the rubbish. Every firm is required to stockpile its debris in fenced off areas at the end of the day. The graveyard shift then removes the rubbish to a compactor bin. "Not only is the place neat and tidy each morning, it has the benefit of cutting out the kind of site vandalism you get with a messy site" says Crane.
>
> "Normally, by the time the operatives have changed into their overalls, sorted out their tools, found the materials and moved everything into place, you don't get any work done before the morning tea break. With this system, everything is ready for them as soon as they arrive."
>
> **That'll do Nicely, Building 12/6/87**

Task completion control

Because of the demand for completion as work progresses there is also a demand for every last nut and bolt to be fixed before the gang leaves the work area. There is no opportunity to return as another gang is occupying the area and anyone returning may damage subsequent work. This is another area of failure in the UK industry; the amount of rework due to damage is over 8% of total value. In the USA it is approximately 1% and in Japan it is unknown. In a component assembly based construction site, replacing unique factory made components because they have been damaged is a waste of time, energy and expense. A disciplined approach is necessary to ensure that:

❑   The work is 100% complete;

❑   The work has been checked and complies with the requirements under the prevailing quality assurance scheme;

❑ The work area has been cleaned and all rubbish removed to the disposal point;

❑ The work area is left safe.

Quality - a 'total view of production performance'

Total Quality Management (TQM) or, more properly, 'right first time', is essential for highly productive projects. Explained simply the approach is to ensure that each step in a sequence of activity is done correctly, at the right time and that there are no gaps or inefficiencies in the sequence. With construction this is difficult because of the one-off nature of the work and the very long and interconnected sequences of activity from initial design to installation and commissioning. This is why the calibre of the people must be so high and why the organizational structure must be carefully designed so that it enables individuals to concentrate on performing their tasks correctly.

It is the specialist sub-contractors who possess the knowledge and expertise and are in the position of directly applying their skills to the project. Thus they can focus on their own requirements for ensuring that they operate efficiently. In practice this means that the design and off site processes must be co-ordinated to ensure that the correct products and components are delivered in the necessary volume, on time and in the order required. Due to the focus on speed of site assembly there is no allowance for error and incompetence. Visible, active and questioning management is required.

TQM is more than a high profile control system it is a way of life. Everyone in the organization must be aware of it and accept the concept as their own. They must take a pride in doing things correctly and not blame others for their own failures. It requires an ability to plan and identify problems and then solve them before they occur. It requires a high level of commitment which is rewarded in turn by the immense satisfaction of seeing the project proceed in a smooth, orderly way.

The practical issues of 'right first time'

The application of these principles is difficult, but the considerable advantages make the effort worthwhile. The frustrations will be at project level, because whilst one or two companies may be implementing TQM, others will not be similarly organized. This is not to say that the effort should not be made in the first place. The contractor and other sub-contractors will soon realize that the company operating TQM is achieving its goals with apparent ease, or at least making considerable demands on others who are not performing.

The major problems occur at the interfaces with the rest of the project. They must be controlled and this requires that they are identified in advance, which will put a considerable onus on the sub-contractor's management to pre-plan and think through the problems. Each one should then be investigated to ensure that the other side of the interface is going to operate correctly and will produce the right result at the right time and that the result will not interfere with site production. The majority of the problems will occur with:

❑ Design teams whose objective is to manipulate the sub-contractor's capabilities to their own ends with little regard for the implications on manufacture, site production or time scales;

❑ Suppliers who have no understanding of the order of construction and thus the sequence of material and component supply. Suppliers that insist on supplying materials with little regard for their future handling on site;

❑ Non TQM sub-contractors who do not appreciate the chaos they may cause through failing to do their work when they should, not finishing it properly or trying to work in the same area as others;

❑ Main contractors who fail to manage the supply of good quality working space and a good working environment;

❑ Contractors who do not keep the site clean or provide adequate materials movement facilities;

❑ Contracts which do not allow the clear definition of responsibilities for each stage of the production process and are negative towards producing efficient operations.

Summary

Productivity must rise from its current low level, if only to ease the current and future problems of craft skill shortages. The biggest barrier to this is the inability to progress work because of interference. The sub-contractors must attempt to control this, first within their own organization, and then through the interfaces. The logical development would be for the sub-contractor to become a vertically integrated organization, thus minimizing the interfaces.

This would mean that the sub-contractor would develop into a trade contractor by assuming complete responsibility for delivering his product.

8

Quality management

"I would like to be the first in the field .... as a marketing tool"
Interview statement

The obligations with regard to quality within most standard forms of contract require no more than that if defective workmanship is found it shall be removed or otherwise remedied.

No "assurance" is offered by way of quality control and other checks: obligations in this respect have to be written specifically into the contract. But times are changing and clients are seeking proper quality assurance, furthermore they do not expect to have to pay more for good quality.

Quality Assurance (QA) schemes are beginning to be used in the industry but there are many questions to be answered. For instance, the following three quotations generally sum up what is frequently felt about the current attempt to apply Q A to building:

'Quality is another piece of paperwork'
'Quality is a stick with which to beat the sub-contractors'
'When we get the quality right - we still get beaten on price'

These opinions aptly describe the general confusion, and illusions, that exist about the subject throughout both the professions and the contractors and clients within the industry. The probable reasons for the confusion are as follows :

❏ The industry's traditional experience of QA is on major projects where safety of operation of the final product has been a major factor i.e. off-shore oil platforms, petro-chemical and nuclear projects. In these situations QA has required extensive paperwork checking procedures as a fail safe guarantee, although it has not necessarily resulted in, or been required to result in, cost-effective quality management;

❏ The standards that describe the elements of a quality management system (ISO 9000/BS 5750) were written with the manufacturing industries in mind. Consequently, most people in the construction industry find it hard to relate to some of the concepts and language it contains, although close examination of both the standards and the design and construction of a building project reveals that the fundamental processes to be controlled are in fact the same;

❏ The confusion may be further compounded by the fact that the standard applies to a situation where both the design and production processes for a product come under a single and continuous responsibility. For instance, car production, where the manufacturer is responsible for both design and assembly. In building, this situation is not generally the case. The building designers have become separated from those who construct, they act as independent professional architects and engineers. Many of the detail designers are also independent;

❏ The industry probably feels that its traditional rather ad-hoc way of going about its business has served it well for many centuries and that the imposition of any formal quality management systems to its processes would only stifle its creative design, limit its flexibility of response to widely diverse and unique projects and not necessarily result in a perceived improvement in building quality;

❏ Many clients view quality assurance schemes as another way by which the construction industry can add additional cost to the project through more checking, more paperwork and more people being involved.

So the confusion and created illusion that the construction industry has about the subject can, at the very best, be a genuine mis-understanding of the philosophy and mechanisms of quality management and, at the very worst, be a smoke screen created in order to resist change because of an erroneous self-image that it is efficient in its current method of practice.

Developing a quality organization

A strategy to create an organization which can produce a high quality product or workmanship must address two issues; the personnel and the system in which they work. The two are inseparable and either are ignored at your peril.

People are a vital component and yet we treat them as if they have no intelligence. Many organizations ask their employees to leave their brains behind outside the office, factory gate or site. Yet ask them and you will find they have many interests outside work They serve on PTAs, local councils, voluntary organizations, play team sports and much more, all of which demand organizational and thinking skills. Are these skills used to your benefit? Why are they not capitalised upon to improve the organization? The majority of people work well when instructed what to do, but they can also see faults, although there is often no incentive to voice their views so that the product can be improved. An organization committed to quality however, constantly seeks from its entire workforce their input to modify, simplify and so improve the product or performance. This attitude demands that those operating the system take it apart piece by piece and put it back together again so as to improve it. An attitude of mutual co-operation exists. This must be a corporate objective which is taken very seriously, otherwise frustration and in many cases obstruction will negate any gains.

Whilst this is often seen as the Japanese approach, it is in fact American in origin. The Japanese have taken it and refined it so that it permeates their attitudes. It can only be achieved by a well motivated workforce, which is very difficult to achieve in construction with its mobile workforce. However, an organization should examine each stage in its operations to see which stage has the most stable workforce and start there in building an attitude of innovation and quality. Large organizations such as Jaguar and Marks and Spencer have shown that they can organize their sub-contractors to deliver to their quality objectives. Admittedly most of their suppliers have a long term relationship and the people are therefore familiar with the necessary attitudes to quality. Stability is important, as are good conditions, information on the current situation, certainty and trust.

The system in which the sub-contractor works, as shown already, is most complex. The sub-contractor must examine his contribution to each part in the process to determine whether his own people can deliver a quality product. The analysis is concerned with an internal explanation of each part of the process, the external constraints on each part of the process and the efficiencies of the transfers between each part. Where the control is internal the quality can be assured. Where the control is external then it is very difficult to assure quality. For example, the interface to the design team is an external control and can only be influenced by an active and positive involvement in order to make sure that the design team does not use the sub-contractor's products and expertise in a way that does not allow him to deliver a quality product. This approach requires that the sub-contractor develops an awareness of the ways in which his activities are controlled and ensures that he has actively taken control of the interface by pre-empting the disruption to his activities. It requires a detailed understanding of every relevant activity in the whole construction process.

The aspects of current practice that militate against the sub-contractor achieving a high quality level of performance are:

❑   That long term relationships, not only within their own organization due to the high mobility of the labour force, but with other organizations which change from project to project, cannot be developed. Therefore feedback and review loop to learn from the latest experience is difficult to develop;

❑   The difficulty of making an input into the early design phases of the project, so the sub-contractor has to rely on an impression of the definitions of others as to what specific requirements he should meet;

❑   The contractual arrangements that put the sub-contractor in an indirect relationship with the building client and the consequential loss of the direct supplier/customer relationship.

These practices create the culture in which it is hard to accept that the fundamentals of quality - the cost effective achievement of all of the clients' requirements through ensuring that non-conformance is prevented - can never be realised. This is a particular problem for the sub-contractor who finds himself at the end of the chain of responsibility, with the least amount of control in the project.

Quality - the practical issues

In order to dispel the confusion and illusion about quality management, each of the three quotations cited at the beginning have been addressed in turn.

'Quality is another piece of paperwork'

Yes, quality assurance systems do involve more paperwork because this guarantees that the work has been checked. It becomes a problem when the items required to be certified are not relevant to the quality of the final product and therefore completing reports on them are seen to be irrelevant. Paperwork will also mount up if the work is not correct before it is checked and a continuous stream of errors and error reports start. This is where management must take responsibility for ensuring that the work is done properly before presenting it for approval. Responsibility for finding mistakes in the work should not be delegated. The paperwork system must be as simple as possible and easy to complete, but also cover all the relevant points. There are often conflicts between the contractor's requirements and the sub-contractor's own system. A good, well conceived system of records is likely to be accepted once it is fully explained. It could also reduce some of the unnecessary paper that already abounds.

A manufacturing company that implemented a quality management system actually reduced the amount of paperwork it used, so that every piece of paper existed only to serve a useful purpose. There is no reason to believe that this cannot occur in a building project.

'Quality is being used as a stick with which to beat the sub-contractor'

Currently, because the subject is still in its infancy, there is an unfortunate tendency to demand QA of the next participant down-line without really knowing what it means for either party. The specialist contractor unfortunately finds himself at the end of the line.

This tendency is either the result of ignorance or is a crude means of self-protection on the part of the participants up-line. A good project quality management system can make as many management demands of the up-line participant as of the down-line participant. In other industries and companies the key to success in implementing cost effective quality management has been the recognition of the customer/supplier relationship within and between different processes. This helps to identify the cause of costly errors and low standards, because it is the defect in the process output of the supplier to the customer that is in error. Once the deficiency in the supplier's process can be identified, the cause of that error can be removed.

Although the specialist contractor, regardless of whether he is sub or trade, is essentially the supplier to the general or management contractor, and ultimately through them to the designer and client, he is also their customer in terms of his need to receive:

❑ Clearly defined and achievable requirements;

❑ project information and management support;

❑   agreed acceptance criteria for his work and;

❑   payment for completed work.

The improvement process should be an ordered, disciplined, management-led focus on quality of work in every department across the whole company. The steps require a constant, controlled management approach to answering these questions:

❑   what is our purpose as a department?   )
)
❑   who are our customers and suppliers?   )
)   "doing the right
❑   what are our customers' requirements?   )   things"
)
❑   where are we not fully meeting them?   )

❑   which is the priority issue?   )
)
❑   what are the root causes of the problem?   )
)
❑   which is the best solution for resolving   )
    the problem?   )   "doing things right"
)
❑   has the problem been eliminated to the   )
    customer's satisfaction?   )
)
❑   have we prevented it recurring?   )
)
❑   which is the next priority issue?   )

❑   when we have removed all the major   )
    problems to customer satisfaction, how   )
    can we operate more efficiently?   )   "continuing to seek
)   improvement"
❑   what are the priority areas for improving   )
    efficiency?   )

Source: How to take part in the Quality Revolution: A Management
            Guide, PA Consulting Group

A quality management system should not be used as a stick, it is much better that it should be used as a carrot for the benefit of all the project participants. It can equally be used by the sub-contractor as against him !

'Even when we get the quality right - we still get beaten on price'

The most difficult aspect about the subject of quality is the need to explain and convince people of its potential for cost saving. This difficulty arises because of a natural tendency to think that quality means raising standards and raising standards means higher costs. The reverse is actually true because meeting requirements right first time every time, can only reduce cost, as well as raising standards, because the sub-contractor is not incurring avoidable costs.

Summary

In conclusion, the specialist, trade or sub-contractor should:

❑ Accept that quality has become an issue because of a failure by the industry to produce an acceptable product. QA systems are only a prescription to attempt a cure; the significant cure is a cultural shift by each and every organization to produce a quality product. A quality organization is a way of life not an ability to fill in forms;

❑ Obtain, with guidance, a thorough understanding of what a Quality Management System could mean for their own business and "specialist" contribution to a building project This is best done with a consultant who will work with the company to increase awareness rather than one who tries to sell a packaged system;

❑ Ensure that the other project participants with whom they will have to work are also practicing the same principles through a project Quality Plan, once their own Quality Management System has been installed in the company. This does not necessarily mean certification to BS 5750. In a building project, any one participant's quality depends on, and will affect, that of another;

❑ Use the company Quality Management System as a means of constantly seeking improvement so that standards are continually being raised at the same time that costs are being reduced. This improvement in quality will raise the organization's standing in the market place with an ability to attract work not necessarily sensitive to price.

9

Safety

The incidence rate of fatal and major injuries in the construction industry is one of the highest amongst all industries in the UK. By the number of fatal injuries construction was the second most dangerous industry on average from 1981 to 1987/88. Unfortunately, as shown in table 9.1 the safety situation in the construction industry has not been improved recently. When the overall number of accidents across all industries is compared by the total number of fatalities and major injuries, construction still ranks as the second most dangerous industry (see figures 9.1 and 9.2).

Deaths and major injuries occur on construction sites all too often, particularly in some specialist working areas. The result of the 18 month 'blitz' by the HSE, which started in mid 1987, has shown the serious safety problems of small sites and, more seriously, over 75% of the problems involved specialist work. During the eight year period 1981 to 1988, 1062 people were killed in the construction industry, 784 were employees, 189 were self-employed and 78 were members of the public (including 21 children). The post war improvement in the reduction of the annual death toll has ceased in recent years. This fact, coupled with the increase in numbers of serious injuries reported to the Health and Safety Executive (HSE) has prompted a number of initiatives designed to bring about improvements, endorsed by the Health and Safety Commission (HSC):

❏ Improved management of construction sites and better control of contractors and sub-contractors' activities.

❏ Improved safety standards within small firms.

INDUSTRY		1983	1984	1985	86-87	87-88
Construction	Number	150	124	139	139	157
Incidence rates(per 100,000)		11.6	9.8	10.5	10.2	10.1
Agriculture & Forestry	Number	64	61	75	66	55
Incidence rates(per 100,000)		8.6	8.8	6.1	8.9	5.8
Energy & Water	Number	54	71	62	37	35
Incidence rates(per 100,000)		7.4	7.9	7.8	5.8	6.9
Manufacturing	Number	135	150	129	115	100
Incidence rates(per 100,000)		2.2	2.7	2.4	2.1	1.9
Service	Number	170	179	214	131	178
Incidence rates(per 100,000)		0.8	0.8	0.7	0.6	0.6
Unclassified	Number	24	18	10	17	-

Table 9.1   Fatal injuries reported to enforcement authorities, including local authorities, analysed by industry. 1983 1987-88
Source: Employment Gazette, Health and Safety Statistics 1986-87

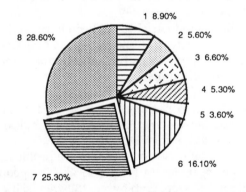

Figure 9.1   Percentage of all fatalities, all industries in UK (87-88)
Source: Employment Gazette, Health and Safety Statistics 1986 - 1987

Notes:
1 Agriculture, forestry and fishing.
2 Energy and water supply.
3. Extraction of mineral ores.
4. Metal goods, engineering and vehicle industries.
5. Other manufacturing industries.
6. Total manufacturing industries.
7. Construction.
8. Service industries.

❑ Improved quality and quantity of safety advice that contractors seek or provide for themselves.

❑ Greater control of high risk activities.

❑ Increased involvement of professional advisers.

❑ Consideration of health and safety at the pre-contract stage.

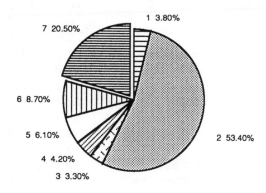

Figure 9.2   Percentage of major injuries all industries in UK (87-88)
Source: Employment Gazette, Health and Safety Statistics 1986 - 1987

Notes:
1. Agriculture, forestry and fishing.
2. Manufacturing
3. Metals manufacturing
4. Chemical industry.
5. Mechanical engineering.
6. Food drink and tobacco manufacturing.
7. Construction.

The following charts and tables analyse the general information and safety problems in the construction industry relating in particular to specialist contractors. Table 9.2 shows that the percentage of private specialist trades has increased in recent years, particularly the employment of operatives working on construction sites, while the absolute number of specialist contractors is decreasing. From the analysis of private specialist contractors, it is apparent that the specialist trades in the UK construction industry are increasing. Specialist contractors' employees account for an average of 44% of total employees amongst private contractors

(October each year)	1982	1983	1984	1985	1986	1987
General builders	213.1	208.6	197.7	181.3	174.8	177.1
Building & civil engineering contractors	102.2	98.4	92.0	84.8	79.3	79.3
Civil engineers	38.6	38.2	38.2	38.0	36.5	37.5
Main trades	353.9	345.2	327.9	304.1	290.3	293.9
Plumbers	18.2	17.8	16.8	14.8	14.5	15.6
Carpenters & joiners	13.9	13.5	13.0	11.1	11.1	11.9
Painters	34.6	34.8	32.9	29.0	28.3	29.1
Roofers	16.6	16.7	17.2	15.9	16.6	16.4
Plasterers	9.5	8.8	8.4	7.1	7.1	7.0
Glaziers	8.9	9.8	10.4	10.2	10.8	13.6
Demolition contractors	2.6	2.6	2.6	2.6	3.0	3.1
Scaffolding specialists	11.8	12.4	12.0	12.7	12.1	12.3
Reinforced concrete specialists	4.2	3.7	3.7	3.5	3.4	3.5
Heating & ventilating engineers	34.4	34.9	33.4	30.0	31.0	30.3
Electrical contractors	49.7	50.1	50.8	52.8	52.4	54.4
Asphalt & tar sprayers	9.2	10.8	10.1	9.2	8.5	8.8
Plant hirers	15.1	15.5	14.8	15.0	13.9	14.0
Flooring contractors	3.9	3.7	3.4	3.5	3.7	3.7
Constructional engineers	7.4	7.4	7.1	7.0	7.8	8.5
Insulating specialists	9.2	8.9	8.6	7.2	7.8	7.7
Suspended ceiling specialists	1.8	1.8	1.7	1.5	1.7	1.9
Floor & wall tiling specialists	2.4	2.4	2.2	2.1	2.1	2.4
Miscellaneous	10.7	11.1	11.2	9.5	8.6	10.3
Specialist trades	264.1	266.7	260.3	244.7	244.4	254.5
All trades	618.0	611.9	588.2	548.8	534.7	548.4
% specialist trades in all trades	43.23	44.08	44.75	45.08	46.20	46.9

Table 9.2   Private contractors - employment of operatives by trade or firm 1980 - 1987 (in 1,000s)
Source: Housing and construction statistics 1977-1987 (information relates to employment by firms on the Department of Environment register)

Table 9.3 shows that the number of reported fatal and major injuries on construction sites is increasing. The percentage of these injuries in relation to the total employees and the self employed is also increasing. According to the HSE, there were, on average, two deaths every week on construction sites between 1981 and 1985, 90% of those deaths could have been prevented and in

70% of the cases, positive action by management could have saved lives. The 157 deaths in 1987/88 was the highest recorded death toll in the 1980s, which may be a warning for the future.

Status of person	1984	1985	1986*	86-87	87-88
Employed					
fatal injuries	100	104	24	99	100
rate per 100,000	9.90	10.44		10.24	10.16
major injuries	2286	2239	579	2546	2574
rate per 100,000	225	225.8		261	259.3
sub-total	2386	2343	603	2645	2674
rate per 100,000	236.24	235.24		273.53	271.75
Self-employed					
fatal injuries	17	22	2	26	41
rate per 100,000	3.7	4.7		5.3	7.6
major injuries	79	113	29	412	450
rate per 100,000	17.03	24.04		84.60	83.03
sub-total	96	135	31	438	491
rate per 100,000	20.73	28.74		89.90	90.63
Public					
fatal injuries	7	13	1	14	16
major injuries	75	78	24	594	587
sub-total	82	91	25	608	603
Total Fatal Injuries	124	139	27	139	157
Total Major Injuries	2440	2430	632	3552	3611

Table 9.3   Injuries to employees and non-employees reported to HSE in construction industry from 1984 - 1988
Source: Health and Safety Statistics 1986-87 and the HSC & HSE Annual Report 1987/88

*   figures for the first quarter of the year. From 1986-87 onwards the year commences April 1

Table 9.3 and figure 9.3 show the more serious situation in which the number of reported fatal injuries and major injuries for the self-employed has increased dramatically over recent years, because of the rapid growth of small firms, self employment and sub-contracting.

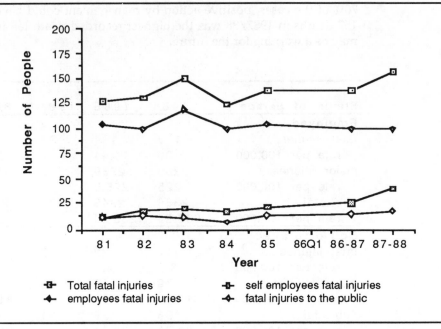

Figure 9.3   Fatal injuries reported to HSE in construction industry from 1981-88

Nearly 80% of the self-employees worked as specialist contractors. *"The lack of safety precautions and poor site management are mainly responsible"* for the deaths.

Table 9.4 shows the actual figures, based on occupation, of the fatal injuries reported to the HSE between 1981 and 1985 and that approximately 50% of those fatalities occurred on small sites (defined as employing less than eleven workers). During the inspection 'blitz', HSE inspectors visited 8,272 sites, saw 10,002 contractors and issued 2,046 prohibition notices stopping work where there was an immediate danger to employees or the public. One in four sites had a major safety problem. Moreover specialist work accounted for a large proportion of the work that was stopped: roof work 20%, scaffolds 28%, excavation 8% and electrics 5% which altogether accounted for 75% of the dangerous work being stopped.

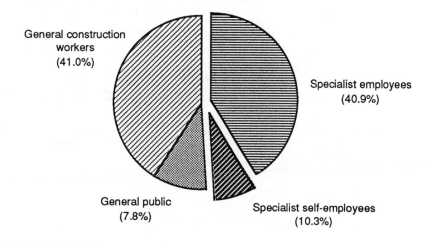

Figure 9.4   Analysis of deaths amongst all construction workers 1981-1985
Source: Blackspot Construction, HSE,1988

Occupation	Employed	Self employed	Total	%
Labourers & civil engineering operatives	193	20	213	31.28
Roofing workers	68	31	99	14.54
Painters	40	13	53	7.78
Drivers	47	4	51	7.49
Demolition workers	43	7	50	7.34
Managerial & professional status	29	20	49	7.20
Carpenters/joiners	30	7	37	5.43
Scaffolders	22	1	23	3.38
Steel erectors	20	2	22	3.23
Bricklayers	15	4	19	2.79
Plumbers/glaziers	10	3	13	1.91
Electrician	7	4	11	1.62
Other/non construction workers	37	4	41	6.02
Total	561	120	681	100.00

Table 9.4   Fatal injuries reported to HSE by occupation 1981-85
Source: Blackspot construction - HSE 1988

The accidents in the above analysis are based on reports of fatal accidents and major injuries to the Health and Safety Executive (HSE). Under the Notification of Accidents and Dangerous Occurrences Regulation (NADOR) self-employed contractors working independently of any other person's direction are exempt from the reporting requirements, so the exact number of fatal and major injuries in the construction industry, and therefore amongst specialist contractors, is probably higher than these data show.

According to the analysis in the HSE's "Blackspot Construction" during 1981 to 1985 and the 'blitz' report last year, the three greatest killers in the specialist contractor's domain were:

- ❑ Maintenance   43%

- ❑ Transport and mobile plant   21%

- ❑ Demolition and dismantling   13%.

Maintenance

Maintenance activities accounted for between 30% and 50% of the total number of construction fatal accidents during 1981 to 1985, with an average over the five years of 43% (see table 9.5).

Many maintenance jobs involve work on roofs and there are three main types of accidents involved in roof work: roof edge falls, falls through fragile materials, and falls from the internal structure of roofs. Together these accounted for approximately 34% of the deaths in maintenance work, while roof workers only accounted for an average of 6.36% of total specialist workers (see table 9.1).

The second largest cause of maintenance fatalities is falls from scaffolding, accounting for 15.87% of the maintenance fatalities. However, the scaffolding specialist workers only make up 4.7% of the total number of specialist workers. The three main types of reported fatalities in falls from scaffolding are: falls from scaffolds and working places, falls from tower scaffolds and falls from scaffolds during dismantling and erection.

The third highest cause of maintenance fatalities is falls from ladders, accounting for 14.92% of the total. Within that category there are two major causes of death: ladders not being securely tied and the lack of reasonable and practicable precautions.

Deaths from electrical hazards account for 6.67% of maintenance deaths. The hazards included: electrical system not isolated, defective portable tools, faulty electrical installations and defective equipment.

Types of deaths	1981	1982	1983	1984	1985
Roof edge falls	5	11	12	7	8
Falls through fragile materials	7	14	13	11	12
Falls from internal structure of roofs		3	1		1
Falls from cradles,suspended scaffolds, bosun's chairs,skips & buckets	3			4	
Falls during the erection of cranes and hoists	3				
Falls during the erection and dismantling of scaffolds & temporary work	4	2	1	4	1
Falls from tower scaffolds	1	4	7		
Falls from scaffolds & working places	7	5	3	5	6
Falls through openings		2	4	1	1
Deaths from ladders & stepladders	9	9	10	11	8
Deaths from falls of less than 2 metres	1				
Deaths from falls on the same level	1		1		1
Insecure loads or unsecured equipment	4	3	2	1	2
Falls of rock or earth from the sides of excavations & tunnels				1	
Collapse of structures or parts of structures		3	2	1	2
Deaths from transport or mobile plant under power	8	8	5	5	6
Deaths from transport or mobile plant not under power		1	1		
Contact with overhead power lines			1		
Electrocution by unearthed or live equipment	4	2	7	4	3
Deaths from fires & explosions	2	1	1	1	
Asphyxiation or drowning	1	2	1	3	1
Accidents not easily classified		1		1	
Total maintenance	57	74	72	60	52
Total construction deaths	140	148	162	137	152
Maintenance % of total deaths	41.21	50.50	44.94	44.29	34.71

Table 9.5 Analysis of maintenance deaths 1981-85
Source: Blackspot construction - HSE 1988

Transport and mobile plant

Between 1981 and 1985, 152 people were killed during construction work when using vehicles and mobile plant. This figure includes non-rail vehicles and mobile plant such as cranes, excavators and dumpers. The deaths occurred during both loading and unloading operations and people falling from plant and vehicles.

This type of accident accounted for 21% of the total deaths during the period, and occurred throughout the whole range of building and civil engineering work. Maintenance and construction of roads, including motorways, accounted for 55 deaths, 36% of the total of 152. The other major activities causing fatalities were pipe laying and sewers (14 deaths) and house construction (20 deaths).

Table 9.6 analyses the accidents by type, showing that being run over or struck by a vehicle moving forward or reversing were the main causes.

Type of accidents	1981	1982	1983	1984	1985
Run over or struck by vehicle moving forward	16	19	10	19	11
Run over or struck by vehicle reversing	5	5	4	4	9
Falls of people	6	4	2		3
Falls of equipment or materials	5	2	2	2	4
Vehicles overturning/stationary	2		3		
Miscellaneous	1	4	1	4	4
Total deaths due to transport	35	34	22	29	31
Total No of deaths reported	140	148	162	137	152
Transport deaths % of total deaths	25.00	22.97	13.58	21.17	20.39

Table 9.6   Analysis of transport and mobile plant deaths 1981-85
Source: Blackspot construction - HSE 1988

Demolition and dismantling

95 people died during demolition and dismantling operations representing nearly 13% of the total of 739 deaths for the period 1981-1985. Of those who died, 47 were specialist demolition workers, and the remainder included a variety of other specialist occupations, like roofing workers, joiners, scaffolders and steeplejacks.

Summary

The analyses referred to so far show that an effective accident prevention and a safety management system on site is crucial and should be built around safety precautions that are both reasonable and practicable and should be coupled with the skill, care and awareness of drivers, plant operators and other personnel.

A recurring theme throughout the statistics is that people are killed during simple, routine work, and so we can conclude that experience in construction work is no safeguard. Experienced workers are just as likely to be killed or injured as trainees, furthermore, certain specialist contractors in their particular types of work, such as roof work, maintenance, transport and demolition, accounted for a greater proportion of fatal and major injuries.

Accidents to specialist contractors' workmen accounted for more than 50% of total fatalities in the construction industry figures, but the number of specialist workers only accounts for around 30% of the total workforce in the total construction industry. Among the specialist contractors, the riskiest activities were roofing work, demolition work, carpentry and joinery work, scaffolding, steel erection, excavation and transport. Figure 9.5 shows a pie chart depiction of the reported fatalities by causation between 1981 and 1985.

Construction accidents must be prevented by effective organization for safety, backed by systems of safe working for all the construction activities, particularly for some high risk specialist work. As written in Guidance on the Safety Policies for the Construction Industry

"........ *duties and responsibilities for health and safety must be allocated and recorded in a clear and logical way with everyone knowing who and what they are responsible for and to whom they are responsible.*"

Construction inspectors have long complained of confusion in situations involving a number of sub-contractors, where safety responsibilities are difficult to define. The main contractor has a responsibility for ensuring that his sub-contractors are working safely, but it was often found that there was a total lack of control on the site. Moreover, whilst main contractors were frequently prosecuted under the general provisions of the Health and Safety at Work Act for the negligence of their sub-contractors, the law is not always helpful in solving the problem. So the safety responsibilities should be more clearly defined in the regulations to reflect more closely the site situation of multiple responsibility. Improved safety guidance and training should be established, particularly where the incidence of accidents is high and largely occur through negligence or a carefree attitude. From the 'blitz' survey, of those in charge of sites, only 13% had a better than basic knowledge of safety, 33% had a very poor level of basic safety knowledge.

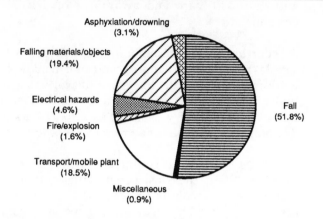

Figure 9.5   Fatalities by causation 1981-85
Source: Blackspot construction - HSE 1988

All operatives have a legal obligation to work safely, although the courts are interpreting this to mean that the contractor's management has a duty to provide safe working conditions for the site. Nevertheless every person should know very clearly his safe pattern of activity, which should be systematically organised at the beginning of the job. Although different specialists have to be trained in different ways, the main points that should be covered are:

❑   Health and Safety at Work Act and allied legislation

❑   Accident prevention, investigation and reporting procedures

❑   Safe working practices on site in relation to different occupations

❑   Occupational protective equipment and clothing

❑   The company's duties and the employee's duties for safety.

Workers on site have to accept the discipline which is implied by the formal establishment of safe methods of working. It is necessary that those in control of the work, at all levels, plan and develop safe systems of work before the job starts.   Active reinforcement of the safety standards and supervision by managers is also important to ensure that workers follow the safety precautions at all stages.   Due account has to be taken of local conditions and circumstances and the attitude, behaviour and trade practices of the worker on site.   There must be a continuous updating of the workers' attitude to safety, during the regular training and retraining process, to ensure that the majority of the trivial errors that lead to accidents do not occur.

The primary responsibility for controlling risks in construction lies with management and the worker on site.   However, a safe site requires the commitment to health and safety of many other people, such as architects, engineers and other professional advisers, as well as main and sub-contractors, safety representatives and safety professionals. Careful design and co-ordination of the work, with particular attention to high risk specialist work, can reduce the overall risks.

Better management of sites through detailed pre-site planning with all who are to be involved in the job is needed in order to improve the general level of safety on sites, especially on small sites.   As has been pointed out, 'more than half of the reported fatalities happened on small construction sites'.   Because more serious safety problems exist in small construction firms and by self-employed operatives, the HSE intends to improve the work on safety standards for small firms.

As an expert remarked:

> *"In the construction industry, at site level, those who create the risks may well be those who work with them, and directly suffer the consequences of error or oversight".*

Whilst this may be true and many accidents could be avoided, there is an unavoidable duty on everyone to guarantee the safety of everyone else involved in a construction project.

10

Future training needs

The last two years has seen increasing shortages of specific skills in certain areas within the construction industry. Whilst the particular cause of these problems may be short lived, already in 1989 there is a 30%+ downturn in housing with a virtual stop in new housing starts in some locations, there is a larger problem looming on the horizon. The demographic time bomb of the next 5 years when the number of school leavers will fall by 25% between now and 1992, coupled with the ageing of the existing workforce, will combine to reduce the available pool of resources from which to obtain the workforce. There will also be stiff competition for more attractive jobs as well as the need to upgrade the quality of the entry to cope with the new technology. The major problem is to maintain a competent and qualified workforce.

The main reason for the skills shortfall is the lack of training during the downturn in construction activity in the early 1980s. Figure 10.1 shows the percentage of trainees per annum of the total workforce, a workforce that is declining. Skilled labour left the industry during the period 1976 to 1987 (as shown in figures 10.2 and 10.3) an estimated 300,000 craft and operative jobs were lost in this period. In 1976 skilled specialist workers accounted for 27.4% of total employees this figure was reduced to 23.9% of total employees by 1985. However, the demand is there, for example, among private contractors in 1976, specialist trades accounted for 34% of the total trades whilst in 1986 they accounted for 44%. This transfer of demand also reflects the changing pattern of demand from the public sector to the private sector.

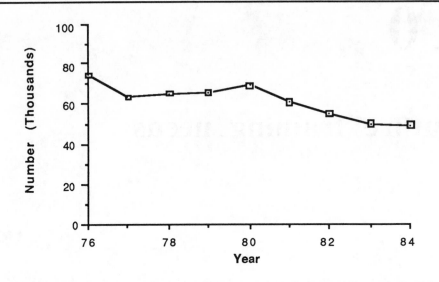

Figure 10.1   Trainees operatives registered with the CITB 1976 - 1984
Source: Housing & Construction statistics, DoE

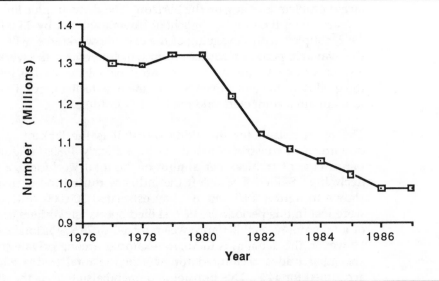

Figure 10.2   Employees in employment in the construction industry
Source: Housing & Construction statistics, DoE

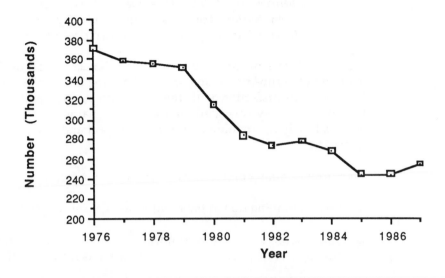

Figure 10.3 Specialist workers (excluding trainees) 1976 - 1987
Source: Housing & Construction statistics, DoE

There are ways in which the problem may be mitigated, but these may not provide the total solution. For example, labour could be imported from Europe as there will shortly be no barriers to movement within the EC. This would probably be per project rather than long term although some European sub-contractors could, and are, forming permanent bases in the UK. However, this also depends on the relative state of the construction industries in each member country as the problem of skills shortages will hit many EC members virtually simultaneously. The same problem seems to be appearing in Japan and the United States.

The UK has been attempting to provide sufficient trained operatives through the Construction Industry Training Board (CITB) but this is currently under review as part of the Government's analysis and review of public expenditure. There is a large debate underway centred around the retention of the CITB. Its main advantage is that it has a unifying role in a diffracted industry. However, it is seen in many quarters as not serving the needs of the industry as it does not provide training for many specialized sectors nor has it stemmed the decline in number of apprenticeships. It is reported, (Building 26 May 1989) that the Electrical Contractors Association (ECA) is about to withdraw from the CITB because it:

> *"..........believes the CITB views the electrical sector as a subordinate organization within the construction industry and because of this, refuses to target training skills to the electrical sector."*

This would be a major blow to the CITB, but is typical of the feelings which abound in the sub-contracting sector. Since the training levy was imposed there has been dissatisfaction with the CITB bureaucracy and complaints that too little of the money was spent on training. What is clear is that if there is to be a central body for training then it must respect and respond to sector and local demand.

The current training problems have arisen for four reasons:

❏ Lack of training due to the rapid growth of self employment

❏ Fewer large firms are committed to training because they no longer employ labour directly and prefer to work as management contractors.

❏ Small firms have to bear more responsibility for training, but they have fewer resources to cope with the changes in the market. Among 88,000 construction firms with more than two employees there were 91% who employed less than 25 employees in 1982. Four years later the proportion had increased to 95%. It is very difficult to see how this vast number of very small firms can take an individual responsibility for training.

❏ The small sub-contractor does not have sufficient skilled operatives to supervise and teach apprentices. This is shown by the fact that the majority of apprentices are trained in the north of the UK where there is still a tradition of main contractors directly employing workers, although this is changing.

The problem is not confined to operatives, there is also the growing need for competent managers within sub-contracting due to the increasing complexity and risk they are carrying. The traditional route to becoming a foreman from the craft base is still needed but with increasing technical content. However, modern demands are also requiring technocrats who are technically and managerially competent to manage the business in this rapidly changing environment.

Operative recruitment and training

There are three measures which need to be considered if the industry is to cope with the problem implicit in the future trends. First, the need for labour could be reduced. Designers could become much more conscious of the link between design and construction and their effect on productivity. Equally sub-contractors, when offering design and construction packages should consider the labour component very carefully and attempt to minimize it by increasing the construction efficiency of the design. Another way is to increase the proportion of off site assembly carried out in more productive factory conditions.

Secondly, the existing workforce must be nurtured and cared for. Every time the worker is made redundant because of a termination of a site or moved to another location there is the possibility that he may be lost to alternative more attractive and stable employment. These opportunities are going to become increasingly attractive as everyone scrambles for trained, competent workers. The industry must make itself less casual in the treatment of its workers. There will be a sharp reduction in the pool of trained workers and the conventional poaching from one employer to another will only cause a rapid escalation in costs. The workforce must be paid properly, the working rule agreements do not recognise the realities of the industry today, they still reinforce the traditional worker/staff divide. The dispute by the steel erectors in March and April this year (Building, 28/4/89) indicated, by the demands they were making, their need to be placed onto a more modern footing with a wage structure similar to that of employees in other manufacturing industries. The demands include: a high hourly rate instead of a low rate plus arbitrary bonus payments; paid meal breaks; full holiday pay; and realistic payments for working away from home. In the end the issue was fudged and an opportunity to rationalize the pay structure was lost.

The company must also work hard to retain the worker by making him part of the team by constantly upgrading his skills through a pattern of retraining. Some examples show what some trade contractors already do. Gartners of West Germany retrain their whole workforce every two years so that they can keep abreast of the new technology and working practice. The supervisors are retrained every six months as well as returning to the factory to plan each installation with the design and production team. Smallman's, the steel erecting firm, retrained welders to undertake stud welding because of a high failure rate when using general operatives. The welders were deemed to have a better attitude and be easier to train than unskilled people for the task. Training can be either in house or, as in the case of Smallmans, undertaken by the equipment supplier. These examples also indicate two approaches one long term the other short term. One company operates internationally to keep a balanced workload the other locally and thus subject to the vagaries of local conditions, both are attempting to improve the multi-skill capabilities of their core workforce.

Thirdly, the companies have to obtain sufficient new recruits to satisfy their future demand. Many major companies in other industries, banking, insurance and the services sector are already gearing up their recruiting so that they will not notice the shortfall. They are recruiting from a young and mature female workforce, ethnic minorities and the previously unemployable. They are offering training and child care facilities, anything in fact to attract sufficient people. The construction industry already employs people from the near unemployable section and will find attracting a proportion of mature women extremely difficult for site work. How many sites can operate a creche safely? The problem is therefore, to attract enough people from a diminishing pool of male and young women entrants. This will be difficult as the entry standards will increasingly be raised to obtain the required intellectual level to cope with the changing technology. It is to be hoped that the schools can improve on the current 40%, who leave with less than the acceptable levels of literacy. There is a need to increase the pool from which everyone will draw.

At present, construction is not even perceived as an industry let alone one which offers careers to school leavers. This is the first obstacle to be overcome. The second obstacle is to provide a pattern of opportunity which rivals that of other industries and that pays as well as they do. Conventional approaches, based on apprenticeships, do not seem to work. Existing training schemes have a very high fall out rate with, for example, only 40% of services trades apprentices achieving the final grade after 4 to 5 years (Gann, 1988). Criteria, based on time served, no longer suffice, the industry requires a highly competent workforce quickly. The measures should be based on competence and rewarded accordingly not based on some 19th century anachronistic arrangement. The ladder of opportunity shown in figure 10.4 is put forward for consideration. It is based on recognizable stages of competency related to the tasks that are required to be performed. Not all tasks need a craftsman's skills and are more related to the specifics of the specialization. Equally, a quickly gained initial competency allows a reasonable pay level to be awarded and allows construction to compete on equal terms with other industries for recruits.

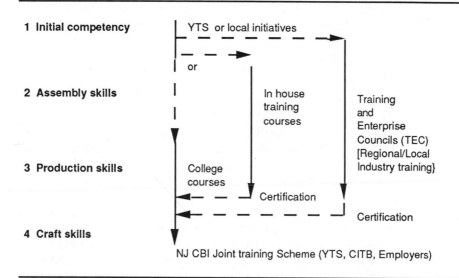

1 **Initial competency**

2 **Assembly skills**

3 **Production skills**

4 **Craft skills**

Figure 10.4 Proposed ladder of opportunity for Operative entrants to the craft skills

Initial competency

If a reasonable level of bricklaying skills can be given to self build groups in 8 weeks, then it is not unrealistic to achieve this for all basic trades. This gives the operative a skill level which enables them to earn a realistic and competitive wage. This is what YTS should be aiming at.

Assembly skills

With the growth of specialized components and off site prefabrication the site assembly skills are changing and are product oriented. This is not de-skilling the industry but the recognition of the way construction and has changed and what training is needed. Many of the fixing techniques are simple: nailing, screwing, gluing, bolting and push fit, in fact all within the competence of any DIY enthusiast.

Production skills

Once competence has been shown at the preceding skill levels then advance into the traditional craft certificate skills can commence in terms of increasing the level of competence and the range of knowledge and ability within the specialization in order to achieve a good level of all round ability in the trade or product.

Craftsman

This would be the final step for those who have shown ability in the preceding stages. It is the equivalent of the German *'Meister'* ie a tradesman of the highest quality with the ability to lead and manage others. The qualification is set at a higher level than advanced craft and includes engineering and management skills.

The objection to these proposals are that it would produce, in the early stages, a potentially half trained workforce because, after the initial training, the operatives would see that they have sufficient skill to get by and start earning a decent income. If the existing pay structure is retained then that will be the case. Should training be done on the cheap? It is an investment for the future.

Already steps are being taken by the BEC and the unions to explore the linkage between pay, recruitment and training, see box below. These steps follow a logical progression to meet the two aims of providing a well paid career structure which provides people able to meet the industry's needs. Each specialist will have to determine how best to provide the training but students increasingly head for the organizations with a well considered training programme. The educational institutions must rethink their approach. At the moment there are too many agencies producing courses which do not offer a coherent pattern to the newcomer. This may be a role for the sub-contractor's umbrella bodies to act as the catalyst to bring together a pattern of need, the Government's agencies, educational establishments and certification bodies. The ECA certainly thinks it can do it.

Management recruitment and training

Production management needs are changing rapidly. Construction management, management fee and fast track construction methods are shaping the structure of the progressive sub-contractors. It has been a painful process for many of them as they have come to terms with the management requirements. Previously it was sufficient to put on site the operative team with a charge hand or supervisor. The contractor would largely take over the team and manage it himself as if they were his own men. The new requirement is that the sub-contractor manages the whole task, including planning his operations, in conjunction with the other trades, to ensure that the increasingly tight production targets are met. The basis of these changes are shown in figures 10.5 and 10.6. Figure 10.5 shows the management structure under a conventional contract where the sub-contractor's management is at level 2 with visiting help at level 3. All of the major managerial tasks are done at levels above by the contractor.

Details of the revolutionary proposals to reform wages and training emerged this week. The reforms, if implemented, will constitute the biggest shake up of employment practices since the early 1920s. Every clause of the 65-year old Working Rule Agreement is being reviewed by the Building Employers' Confederation. In addition the whole structure of training is under scrutiny. The revisions include:

❑ Allowing for a new workforce of "multi-skilled" operatives, adept in the new assembly operations now needed on site, to be trained by a series of short courses.

❑ Ending the restrictions on working hours.

❑ Replacing the archaic structure of pay rates with wages that reflect present day skills and sub-contractor rates.

Building   17/2/89

A properly functioning organization must have all of these levels present within the authority and control system. They must also be aligned so that authority and responsibilities are clear and unambiguous although each higher managerial level is capable of handling increasing levels of ambiguity. The formal relationships must also not inhibit the authority and responsibilities.

The concept of levels is very useful in studying complex organizations in order to determine the type of decision making skills that are necessary and thus the type of person needed to do the job properly. There are six levels of decision making within the hierarchy of production management (Jacques, 1978), that are of importance in this context. As one moves from level 1 to level 6 the timespan of the work (that is the period between making a decision and seeing the practical results) gets greater and the decisions become more complex. The key personal skill which differentiates between the levels is the ability of the individual to abstract and conceptualize the type of work in his own and lower levels. This is more clearly expressed by looking at the levels which occur in a project organization:

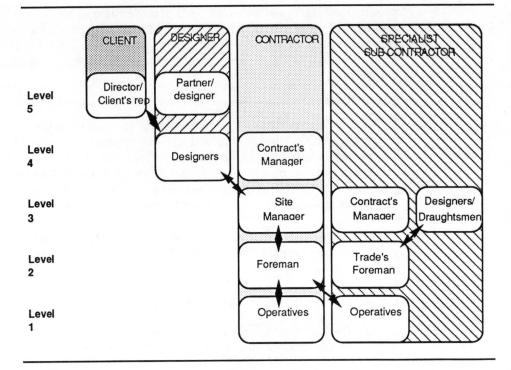

Figure 10.5   Management tasks in a JCT form of contract

Level 1   Direct construction work which requires appropriate assembly knowledge and skills but involves no management responsibility.

Level 2   First line management or a foreman of a team of construction workers. The foreman manages the task allocated to his team.

Level 3   Site management which involves selecting methods of working for several construction teams and ensuring that they have the resources they require to complete the tasks.

Level 4   Construction management which involves managing the site organization including selecting the methods of work and adjusting them as circumstances require.

Level 5   Project director which involves the comprehensive management of all parts which together form the large scale construction project, including shaping their operating environment and creating the policies and systems that they will use.

Level 6   Managing director who initiates the creation of the total business represented by the project.

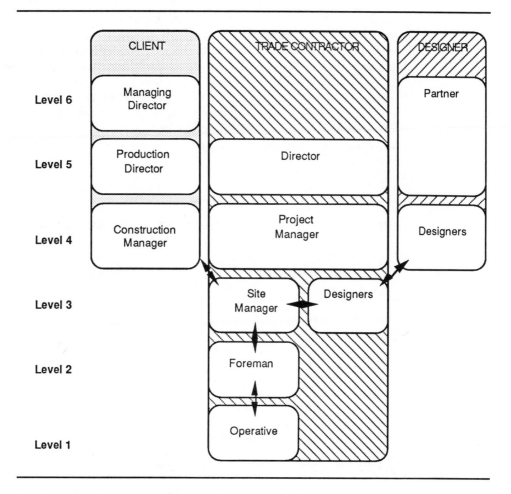

Figure 10.6   Management requirements in a construction management contract

As can be seen in figure 10.6 there is a change from sub-contractor to trade contractor to indicate the upgrading in the role and responsibility. The trade contractor is being asked to provide an active level of management expertise on site at least two levels higher than normal. The management skills are of a higher order as the specialist is responsible for planning and co-ordinating the on and off site work as well as ensuring that the design information is being produced. It is not unusual for directors to be on site as site managers. Obviously this can only be sustained for a limited period and so sub-contractors must start training for this changing need.

A long term strategy to develop the range of management required should be evolved by the various sectors in the industry.   As shown in figure 10.7, a system based upon technical capability supplemented by management skills is necessary.   There is still a need to develop the level two management from the trade base, but the levels three and four are more easily supplied by the technical colleges and universities.   However, it will require a radical new approach by sub-contractors to overcome the attractions of working for the main contractors.   A re-occurring complaint is that sub-contractors cannot deal with sophisticated technical problems at the senior level, which means that they must recruit and promote well qualified engineers through their management structure.

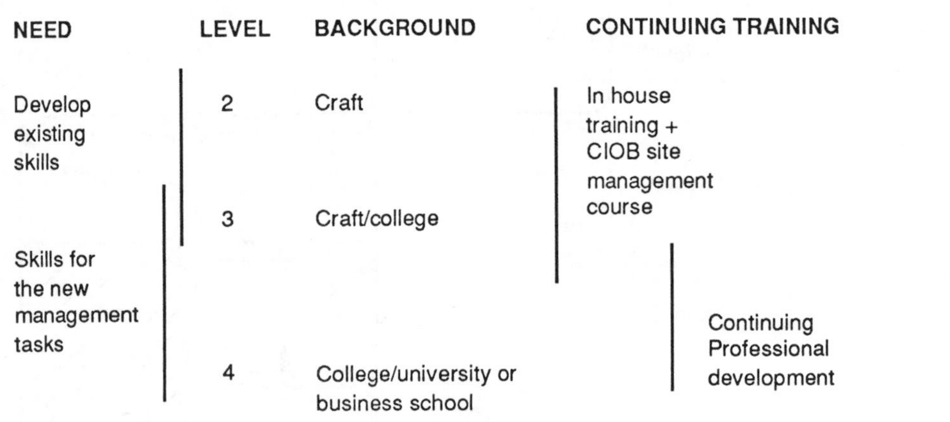

NEED	LEVEL	BACKGROUND	CONTINUING TRAINING
Develop existing skills	2	Craft	In house training + CIOB site management course
	3	Craft/college	
Skills for the new management tasks			
	4	College/university or business school	Continuing Professional development

Figure 10.7   Management training requirements

Summary

The plea for more training on its own will produce only marginal results. There is an obvious need for a trained and competent workforce with an estimated need for between 70000 to 940000 new recruits, depending on the forecaster, by the year 2001.   The intense competition for personnel will mean that the industry will only gain these new recruits if it can provide an attractive, well paid career structure and adequate training.   The majority of sub-contracting firms are too small to provide the basic facilities required by these new people and the current providers of these facilities are reducing their own labour forces and thus the facilities. A slow downward spiral is seemingly inevitable. There is thus a case for the initial skills acquisition to be solely in the school, college or training centre, because only when the operative has reached a certain level of competence will he be paid properly. If this were coupled with skills testing and certification in accordance with the structure outlined above a new type of workforce would start to emerge.

The existing craft skills level needs to be upgraded in technical and managerial content for the new responsibilities. Equally a new breed of management, based upon a higher level of intellectual capability is required to cope with the new levels of planning and organization. Increasingly, graduate engineers will be required for management, particularly if the organization is in competition with European and North American companies.

11

Research and development

In a static industry there would be no need for research and development (R & D). The construction industry does not fall into that category and more change will occur over the next decade. These changes, some of them structural, arise through a shift in emphasis from what has been for centuries a craft based industry, to a modern technology based industry. Many new materials have replaced traditional materials and new forms of procurement have replaced conventional procurement methods. Automation in construction has also increased and the extent of the engineering and environmental services in finished buildings has increased dramatically over the last two decades.

Clients have not been slow to demand a better deal from the industry. Whilst change is still largely evolutionary, the pace at which it occurs has accelerated dramatically. Those who work in the industry need to rapidly adapt their attitudes in order to improve their uptake of technical and managerial innovation.

The requirement today is for an adaptable industry, that is, one that can change to suit the many different demands being placed upon it by its clients. Finding out why something has not worked and ensuring that the fault does not occur again is a basic R & D activity, this applies equally to methods and management as well as products. Effective R & D is thus needed at all levels in the specialist and trade contractors organization.

It has become fashionable to complain that the industry, and the specialist and trade contractors in particular, take little or no interest in R & D. The same can also be said of professional consultancies. So, are there perhaps valid reasons why specialist and trade contractors, do not undertake R & D? It might help to look first at where R & D is carried out and to then consider the circumstances that influence this activity.

In the construction industry, R & D is most evident in the science of materials development and component manufacturing. This is largely undertaken by the major materials manufacturers and suppliers. The government has had an historical involvement in basic research, but it does not seem interested in continuing its funding for, major, long term R & D. While comparisons with other countries in Europe and Japan might put forward a persuasive argument in favour of an increased injection of government funds, the UK Government is unlikely to be impressed. There is, therefore, little point in arguing that as some 40% of all construction work is for the public sector, the Government should "pay its way". It is probably the case that the government is determined to reverse the situation that it inherited ten years ago. When the public sector subsidised private sector R & D. It is imperative that the industry recognises this fact and develops its own strategy for conducting R & D.

The need for research and development

There is a relationship between R & D and business. New businesses start because they have a new idea, which is then developed over time as the business grows. The growth of specialist contracting has shown that this process is alive in the construction industry. The relationship between research and the market can be broadly divided into the following three categories:

New markets

In this case new inventions have generated an opportunity for product development. Examples over recent history have been: glass filament which spawned glass reinforced concrete and thermal insulation; plastics which have changed plumbing and interior design with plastic laminates, and particle board which has revolutionized fitted furniture. It must, of course, be remembered that all these products have taken years to develop.

Market niche

Many companies, particularly small ones, have used this technique to create their own business opportunity. They have taken a product or service and modified it to perform in different ways so that they can offer an advantage which their competitors cannot. Float glass is a large scale example of this, where the basic process of glass making was rethought to give a competitive edge. It is now a leading method, having been licensed all over the world. The same is happening, but on a smaller scale with the application of computers, particularly in products for controls of services and lifts. Here the products are usually small but highly sophisticated, therefore there is a spectrum of opportunity.

Market led innovation

This is the area in which most specialists find themselves. It is where designers take the existing expertise/product/capability and expand it in new directions in order to satisfy the realisation of their own design needs. Whether the result is developed into a new product or market relies upon the specialist. In many cases the changes are so subtle that it is hard to recognise that something new has been created and the market opportunity is often lost. The really successful companies observe this activity and capitalise on it straight away. IBM for example sits with its customers in order to develop new ideas of ways in which its computers can be used they then markets the new ideas in terms of what can now be achieved with its machines, thus selling more and more computers. Therefore listening and learning from the customer is vital and the most obvious practice in the construction industry.

What is R & D?

R & D represents a spectrum of activity centred upon new discoveries of things and innovation. There are steps in this process and the structure and objectives of a company will determine which of the steps are applicable to it. However, it is incumbent upon all organizations to embrace one or other of the steps:

❑ Fundamental or basic research - the process of scientific discovery - is the beginning. Most major new discoveries have been made by chance. Many of the subsequent applications have been in unrelated fields. Thus, a new material which may well have its origins in fundamental research may have no immediate practical application. This is why it is difficult for organizations to contemplate fundamental research, because whilst it may be of benefit to mankind, it is not of direct benefit to the company. This work is largely funded through altruism.

❑ Applied research begins when a potential application is found for a discovery. A new material might find its way into anything from the human body to an office building. The nature of the material will clearly be different in some way and this is what applied research is about. Novel procedures devised for other fields could also find an application in construction.

❏ Development is the final phase in the sequence of R & D activity. It takes the form of any pursuit which seizes upon the results of fundamental and applied research and which is directed towards the design and introduction of useful materials, processes and systems, or the improvement of existing ones. Taking an idea and making it work is crucial to successful R & D, and is an aspect of R & D that is often underestimated. The effort to develop the basic idea or concept into a usable product could be ten times greater than the initial research effort or more.

Of course, R & D is not only concerned about tangible materials or products, it is also concerned with innovation in methods and practices. Increasing management efficiency is a good place to start. The 'acid test' is whether or not the results of R & D increases productivity and gives decreasing operating costs. These are likely to be the more compelling reasons for adopting a new approach.

Methods of Research and Development

It seems apparent that a large R & D department offers only limited success. Its existence encourages the rest of the organization to assume that it is the only one to do the research and to have new ideas. Obviously some types of research need laboratories and fixed equipment, but the majority need an organization willing to listen to new ideas.

Skunk works

The most innovative companies create an environment which encourages small scale, small team working: 'skunk works' (Peters and Austin, 1985). This concept enables anyone to be creative. These teams often work alone, bootlegging the resources that they need, often with the benign acceptance of the company. A more formal approach is that of "task teams", but here size is an important criterion and once the group is larger than about fifteen it tends to lose sight of the objective and lose its spontaneity. The objective is to foster initiative, develop the idea quickly and get the product out into the market place.

Warm armpit marketing

The market place is the most active incentive for R & D. In construction, the creative designers are the market place and it is they who continuously challenge the abilities of the specialist. Instead of regarding it as a nuisance it must be capitalised upon. This is called 'warm armpit' marketing. It involves sitting with the customer and developing your specialism with him. Each new opportunity must then be exploited to create a new market. This can be done by encouraging those who sit with the customer to be intrepreneurs (Pinchot, 1985). In this way if they notice a potential new idea, instead of someone else in the company being given the task of developing it, the person who noticed the opportunity is given the resources to develop the new business. To make this system work they must be given the same incentive and risk sharing as if they were starting their own business.

Creative swiping

Many ideas are taken from other sources and developed into new products, capabilities or services. The Far East nations are masters at this and have created very successful manufacturing industries from it. However, it is not to be confined to them. It can also be done to improve the construction industry here. An example will indicate how practice in the US have been observed, modified and increased productivity in the UK. On the Broadgate project a party from the steel erection sub-contractor and the client visited sites in the US and saw how steel was being handled there. They saw that instead of chains being used to lift pieces of steel, clamps were used, which were much safer and quicker. When they returned they experimented with similar equipment, adapted the technique and were eventually lifting three beams, in one lift, spaced so that the three pieces could be fixed simultaneously. Further studies are underway to look at the practice in other countries to achieve even better performance. These researches are not expensive and are, therefore, very cost effective.

Summary

R & D is not just for the big companies, it is for all companies. Whatever the business there are always ways in which it can be improved. Small is ideal when it comes to innovation, if the opportunities are recognised and the originators fostered.

12

International sub-contracting practice

Over the last ten years the UK construction industry has felt the impact of imported construction practices. Initially these were imported from the USA, first with new management ideas, then with designers and now with sub-contractors. Japan was perceived as a threat but, as yet, it is only their contractors acting as developers and project managers who have made any positive moves. Finally there is the European dimension, accelerated by the potential impact of harmonization in 1992. European sub-contractors in the specialized fields of external cladding and lifts have worked in the UK for many years, but suppliers of basic components such as structural steel frames and precast concrete are now also in evidence.

The changes have been brought about to an extent by individual companies seeking out market opportunities in the UK, but more by designers and clients being uninhibited in seeking components and suppliers from elsewhere to satisfy their aesthetic and production needs, whilst the UK manufacturing base remains essentially staid and static. The foreign companies have found it a relatively simple operation to supply the UK market because of their way of working in their home market. This chapter looks at the practices in those countries which will have the most impact upon the UK component supply industry in the medium term future.

The construction process diagram developed in chapter 3 has been used to show the general principles of the relationships and responsibilities of the most common form of contracting in the respective countries. This will allow a a comparison to be made with the practice in the UK and help show how sub-contracting around the world shares many similarities, but also where the significant cultural and practice differences lie. The same care as before must be used in interpreting the significance in the diagrams.

The USA

The United States of America has the Western world's largest construction industry, spread over an enormous area with many regional variations, leading to difficulties in generalisation. However, there are significant common characteristics which have shaped the nature of the building sector. In macro terms the economy is pure capitalist with a relatively small proportion from the public sector; private capital dominates, pressure to ensure a rapid return on investment, either financially or in additional production output. In micro terms there is a strict division between union and non-union contractors and sites; highly flexible labour employment policies; virtually total sub-contracting; high wage rates (compared to the UK); specialization based on union negotiated demarcations. This has led to a proficient sub-contracting industry operating in a highly competitive market Companies have to be very efficient to survive. On some projects there are between 10 and 15 specialist contractors bidding for the same work package.

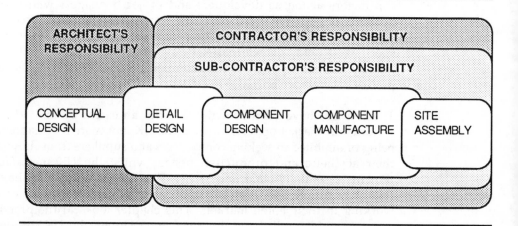

Figure 12.1 The responsibilities within a lump sum contract in the USA

The buildings are characterized by a high degree of repetition and minimized controls on aesthetics, with the components drawn from a standardized, readily available set. US construction is based upon standardisation, repetition and off the shelf components. Their architects are sometimes referred to as "Sweets Catalogue designers" (Sweets Catalogue is similar to the Barbour Index, containing complete catalogues from leading manufacturers)

The sub-contractor is invited to bid by the contractor and visits the contractor's office to do all his measurements of the drawings. There are no bills of quantity or quantity surveyors. The estimator prepares the bid in accordance with the drawings to meet the performance specification. He has tremendous freedom to adapt his product/skill to enable him to lower his price to the most competitive level. This requires that the specialist sub-contractor "trades off" between the design, production and performance of his product to maximize off and on site production and therefore enhance his competitive position. Engineering and production demands tend to dominate at the detailed level and the sub-contractor is responsible for producing a high proportion of the detailed design information. Its production is co-ordinated by the contractor. The design team have fourteen days to review the drawings to determine compliance with the performance specification and will issue one of the following approvals: A - accepted, B - passed at own risk, C - rejected. The onus is on the sub-contractor and the contractor to ensure cross sub-contractor co-ordination of the details.

The main difference between UK and US specialist contractors is in the attitude and in their management abilities. America works on the 'can do and KISS' (keep it simple - stupid!) principles. The competition is fierce, but the rewards for a successful project are good. Contractual claims do occur and things go wrong, but generally projects are completed on time. The job is designed before work starts on site and when the architect changes his mind by issuing a change order, the cost and time penalties, for the client, are high. The management team is better equipped for the job in the US. Specialist contractors are able to recruit good staff by paying good salaries. The companies recognise that they are on their own with minimal facilities provided by the general contractor. Hence they are geared up to survive with the minimum of input from the main contractor. The principle of repeat business is strong, but the lowest bid generally wins the project.

The sub-contractor's site management is of a very high quality and carries total responsibility for the complete package of work and its support services. As time is of the essence he can use the flexible employment system to absorb workload variation. If necessary he can hire and fire part way through the day thus minimizing the high labour costs of surplus labour. The site activities are supported by a very efficient off site materials and components supply infrastructure which in many cases can give same day or next day delivery of the standardized components.

The overall system is geared to high productivity by placing the total responsibility for production, (the integration of design, supply and erection) in the sub-contractor's hands, co-ordinated in detail by the contractor. US specialist contractors already have a significant presence in the UK and work on many of the larger projects in London. It is likely that many more will be seeking work in Europe and probably use the UK for their operational base because of the existing base of US clients and consultants.

Japan

The Japanese construction industry is structured to give a high quality building on time. Being within an earthquake zone the building technology is either very flexible, as in housing, or very strong, as in large buildings using mainly structural steel. It is a very stable industry with the six largest contractors maintaining a constant market share for many years. As with all construction industries its workload is volatile and the flexibility thus created is absorbed by the sub-contractors. Most work is sub-contracted; nonetheless the main contractors are very big organizations, largely because they employ a huge design and administrative staff. Instead of the technical expertise being held and developed by the specialist contractors, it is held by the main contractor. (figure 12.2).

It is interesting to compare the profitability of Japanese contractors with that of comparable UK contractors. Figure 12.3 shows the comparison between UK and Japanese contractors of profitability before tax as a percentage of turnover. A word of caution must be expressed about using any profit figures when comparing companies in different countries due to the different accounting conventions. However, in both countries there is a strong similarity in the margins, bearing in mind that both of these data sets are prepared from consolidated accounts, which include property, housing and other interests. The figures in table 12.1 show the profitability of selected Japanese specialist contracting companies. The companies have been divided by activity into three groups. Comparison can be made with the similar data for the UK mechanical services contractors in Chapter 1 (figure 1.12).

Contractors work for the same clients for many years, perhaps decades, and projects are often awarded on the basis of a negotiated price. The Japanese business culture is very different to that in the UK. Working relationships are based on trust and are built up over many years. Most Japanese contractors have a strong commitment to completing projects on time. What characterises all Japanese companies is the meticulous attention to planning prior to any work starting.

Specialist Contractors	¥ Million		
	Sale	Operating Profit	%
Electrical Contractors			
Kadeko Co.	360,984	16,491	9.5%
Kinki Electrical Construction	310,994	18,308	10.8%
Tokai Electrical Installation	182,389	7,503	9.1%
Kyushi Denkikoji	141,972	4,288	8.0%
Chugoku Electrical	115,899	7,664	11.6%
Tohoku Electrical Construction	102,145	4,770	9.6%
Nippon Telecomm Construction	92,188	5,049	10.4%
Nippon Densetsu Kogyo Fuyo	79,547	1,085	6.3%
Kyowa Dnsetsu	105,294	5,530	10.2%
NEC Corp. & Construction	2,304,392	87,029	8.7%
Sumitomo Electric Industries	550,115	24,610	9.4%
Shikoku Telecomm Engineering	49,265	1,887	8.8%
Daimei Telecomm Engineering	52,231	2,195	9.2%
ELECTRICAL average			**9.4%**
Mechanical Services Contractors			
Sanki Engineering	137,898	3,757	7.7%
Takasago Thermal Engineering	133,629	3,506	7.6%
ODD Corporation	99,259	925	5.9%
Taikisha Ltd	70,117	2,641	8.7%
Okazaki Kogyo	46,591	1,616	8.4%
Asahi Kogyoshi	48,407	911	6.8%
Sanko Metal Industrial	33,051	1,108	8.3%
Tkyo Denki Komusho Co. Ltd	29,134	2,630	14.0%
MECHANICAL average			**8.4%**
Specialist Engineering Services			
Mitachi Plant Eng'g Construction	164,732	3,444	7.0%
JGC Corp.	220,763	6,852	8.1%
Toyo Engineering Corp.	171,728	220	5.1%
Taihei Dengyo	46,034	3,007	11.5%
Toshiba Engineering & Construction	87,172	2,676	8.0%
PLANT ENGINEERING average			**7.9%**
TOTAL AVERAGE			**8.6%**

Table 12.1 Profitability of specialist contracting companies in the Japanese construction industry. (1988)
Source: Japanese company handbook Spring 1989,

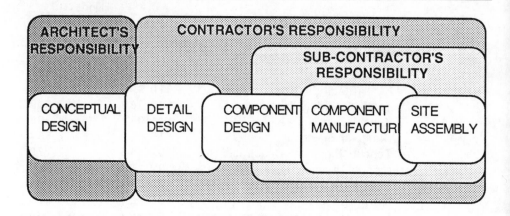

Figure 12.2   Production responsibilities within the Japanese form of contract

Nearly all building work in Japan is undertaken by the specialist trade contractors, who enjoy a paternalistic relationship with the general contractor. They depend on their 'father' contractors for future work. Many of the specialists will have worked for particular major contractors for many years, and in many cases they will only work for one contractor.

Japanese specialist trade contractors fall into two categories. Firstly there are the installation arms of the major electrical, mechanical, component and equipment manufacturing companies such as Toshiba, Mitsubishi and Hitachi. They also get involved in the design and manufacture of a wide range of products from turbines to curtain walling. Secondly there are the independent specialist trade contractors ranging in size from the small labour only gang to the large firm. Most specialist trade work is labour only. Each sub-contractor maintains a core workforce which is supplemented as needed by additional sub-contracted teams, often based on the family unit. It is not uncommon to find husband, wife and children working as a team in a part of the building. The core workforce will be trained in the new construction techniques.

The special relationship between contractor and sub-contractor means that at the bid and award stage the general contractor generally stipulates the contract price, instead of letting the specialist estimate the price for the work. The specialist trusts that the contractor will fairly represent his interests. Contractual relationships are more likely to be based on negotiation than on competition. This means that conflicts simply do not arise regarding payment and claims for additional expense, and even when disputes do arise they are not the subject of litigation. To resort to the courts would be the last resort in a very lengthy negotiation; because of the need to save face and preserve a long term relationship the specialist will negotiate a settlement.

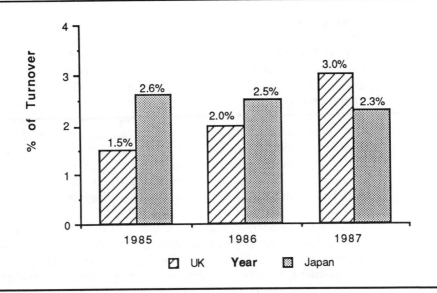

Figure 12.3 Percentage of profits before tax on turnover.   Construction companies in UK & Japan (1985-1987
Source: Key business ratios Third edition 1989 Japanese company accounts handbook spring 1988

Site management is undertaken by the main contractor using a large site management team who are responsible for all materials, plant, planning and organization.   Planning and communication are vital issues with a daily planning meetings at 3pm between the site manager and the key personnel from the sub-contractors. At the start of the shift there is a tool box meeting, here the site manager and team foremen explain the schedule for the day, highlight any major events or problems and generally make sure that everyone is aware of what is going on and their part in it. The workforce is very disciplined due to their cultural background and the need to maintain the highly personal negotiated status and so be able to work on future projects. Payments are negotiated on a weekly basis with the project manager in a bargaining session. There is an absence of quantity surveyors. Quality and time are the ruling disciplines, cost is the variable factor and thus Japanese buildings are relatively more expensive. The pressure of time can be enormous and it is the small and family based sub-contractors who are used to give the required flexibility of resources. The contractor controls the whole design and construction process and manipulates the design and the resources supplied by the sub-contractors in order to guarantee the timely delivery of a quality product.

Contractors monitor the performance of their sub-contractors to ensure that they are sound and their performance is steadily improving. Each sub-contractor is required to report his financial performance and work capacity at the end of each financial year. Twice a year the construction firms' project managers evaluate the sub-contractors and their foremen. Sub-contractors face evaluation of the quality of their work, their ability to complete on time, ideas for cost savings, safety record and quality of management. Foremen are evaluated on their general capability, safety consciousness and cost consciousness. These reports are then reviewed by a sub-contractor evaluation committee of senior staff, and some sub-contractors and foremen are commended for their performance. Any failures of performance are discussed at a senior level and such bad reports influence the amount of work given to sub-contractors.

Japan has also been considering the role of their sub-contractors. They believe that in order to achieve an efficient construction industry the specialist contractors must modernise their operations. They must become more independent, reliable builders and stop operating like manpower agents. The general contractors hope that, instead of merely supplying the number of workers, the specialists will become equal partners.

Western Europe

Surprisingly there are few recorded studies of construction practice in the various European countries. However, the European market is an important area to consider because over the last ten years there has been an increasing penetration of European specialist sub-contractors into the UK, and this is coupled with the advent of harmonization in 1992 and the opening up of opportunities. For the purpose of this section of the study the industries of West Germany and France will be studied as they are probably representative of the alternative construction procurement systems that operate in Europe.

West Germany

The state plays a very large part in the construction industry both in workload and systems. West Germany is a federation of states each having a great deal of autonomy over their own economy through regional parliaments and local tax revenues. There is a tax upon local business and this is used largely for investment in the local and regional infrastructure and townscape. The workload varies from region to region, within each region the majority of organizations, contractors and sub-contractors are locally based to serve the regional. Even the larger companies only rarely interchange personnel between regions. The organizations, therefore, tend to be small and combine together in joint ventures in order to provide sufficient manpower and spread the risk on the medium to large scale projects. Joint venturing is also used to minimize the employment of labour due to the high cost of employment caused by legislation. Labour costs contain a high component of social security, pension and health costs which raises the basic labour rate by 120%.

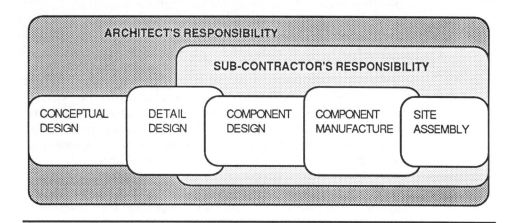

Figure 12.4 Production responsibilities within a West German contract

The design and construction system is primarily within the control of the design team, which may be multi-disciplined or a joint venture. Design is carried out in two stages, firstly, for the overall project concept through to project initiation, and secondly the project procurement and detailed stage. This is called the 'bauliter' system. Under this system the design team divides the project into packages of work. The main package is: ground works, drainage, concrete substructure, waterproofing, and concrete structure; this is usually let to a general contractor who undertakes all the work as well as being responsible for basic site facilities. The second package is the exterior closure system, including the roof, to provide the watertight envelope. After this up to forty separate packages are let and co-ordinated by the 'bauliter'.

The on site co-ordination teams are very small, from one to ten people depending upon the size of the project. Each package contractor has a responsibility to design and manage his work in conjunction with the other packages. This is less of a risk than it first appears because the technology comes from a standardized set of which the sub-contractors are specialists in their own field. There is a trend towards increasing specialization in order to maintain a competitive edge because the market place for conventional skills is intensely competitive. Out of this, particular features have endured: highly engineered systems to build the basic technology, eg Hunnebeker's formwork systems, and high quality technologically sophisticated systems which are world leaders, eg Gartner's external cladding systems.

The overall emphasis is on producing a quality product. Under the standard form of contract (VOB) all the parties, designers, contractors, and trade contractors join together to give a minimum guarantee of two years against design and construction failure. This may be extended to five years in the future. The guarantee is given through the banks to ensure that the insurance will be valid in the event of failure by any party, and when the consortium is broken up after the project. This also has the effect of ensuring that the industry continues to use tried and tested technology. With the emphasis on quality and minimal labour resources, construction time is not particularly quick and cost control is quite poor, with cost overruns up to 20% over project budget. There is a small movement towards project management by external consultants, due more to the complexity of large projects rather than any pressures to bring time and cost under control.

The West German system has recognised the need for flexibility and has been based on sub-contracting with two important features. Firstly, the important packages of structure and envelope must be co-ordinated by one contractor. Secondly, there must be overall project control of design and construction, in this case by the designers.

France

The French construction industry has very similar characteristics to the UK industry in terms of size and mix of work. The gross turnover, at £43.3 billion, is approximately 25% higher than the UK with the proportion of repairs and modernization, at 44.7%, very similar. However, the split of the remainder shows a significant difference. The French have spent heavily in civil engineering. In 1987 they spent £6.8 billion on roads and transportation infrastructure and have plans to spend far more with the investment in the TGV system to give French railways the ability to act as the hub of the European transport system. There has been an extensive investment over the last five years to upgrade the housing stock, 23% of the gross turnover has been in new housing. It is in commercial building that the UK has been investing at a greater rate proportionately, although the French are making significant investments in this area. The majority of commercial investment is concentrated in the Paris regional area, although the towns of Grenoble and Toulouse in the South are the growing centres of the high tech. industry. The government plays a significant part in financing all forms of construction; nearly all civil engineering, a high proportion of housing costs are subsidized by loans as well as mortgage relief and commerce is supported through local and regional aid distributed through local chambers of commerce at favourable rates. However, the French construction industry has yet to reach the level it achieved, in real terms, prior to the 1979 slump.

Building work is organized in a very different way than that in the UK. The French architect has a relatively minor role in that he is only responsible for the conceptual design of the project and for obtaining the necessary planning approvals. On projects over 900,000FF an open design competition must be held. The client is protected by a project based insurance system which, if anything goes wrong, pays for the repairs and thus clients are reported to be satisfied with the service they get from the industry. Art in architecture is, therefore, not constrained, but the architects are not involved in its realization. This is undertaken by either engineering consultants or multi-disciplinary organizations, *Bureau d'etudes techniques* (BET). There is also a management group, *pilotes*, who are responsible for co-ordination. The whole execution process is heavily engineering orientated with the BET, for example, having financial and contractual freedom under the direct control of the client.

Traditionally the construction process has been let on separate trades contracts for all sections of the work. On the larger projects a general contractor may be appointed who would then sub-contract to the trades contractors. There are other combinations of groupings of trade contractors who may join together for one project. But in nearly all situations there is a *pilote* whose role is to organize and co-ordinate the work (see figure 12.5). His role is very similar to a UK construction manager. The trades contractors take total responsibility for their work and are graded by the insurance companies in terms of their capability to undertake certain sizes of project. Below the trades contractors there are the sub-contractors and labour only sub-contractors. The French industry is very competitive and as a consequence of this system there is a polarization of a few very large companies and the rest who are quite small. There are very few medium sized contractors as compared to the UK. Besides the national contractors it is only services trades contractors who are of sufficient size to work across the regions.

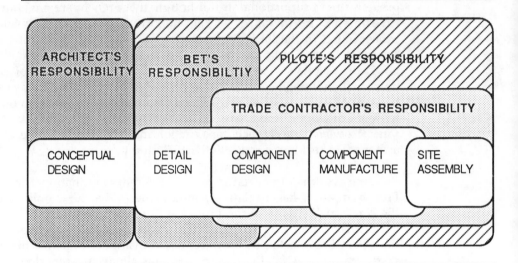

Figure 12.5 Project responsibilities under the conventional French contracting system

Summary

In all of the significant world markets there is extensive use of sub-contracting throughout all stages and levels of a project. A very large proportion of the resolution of the technical detail is in the hands of the sub-contractors, but the risk is limited by the fact that they are working within a limited range of technology.   Each country places great emphasis on there being single responsibility for integrating the sub-contractors' design contribution into the whole design and construction process; in the US and Japan it is the contractor and in Europe the designer. In all cases this has resulted in an over-emphasis on either time, cost or quality. In Europe and Japan quality is the highest priority, although certainty of project time is very important in Japan (although not very quick by international standards). Time is very important in the US, as is cost, but some aspects of quality (as defined in the UK) tends to suffer.

Europe and the US have developed through the demands of their system a high calibre sub-contracting system, heavily oriented to giving high quality, engineered solutions. Their managers are aware of their own need to perform on site which involves ensuring that the total design, procurement, manufacture and assembly cycle is efficient and works in conjunction with the needs and practices of other sub-contractors.   It is not surprising that they appear proficient by comparison with UK sub-contractors.

As shown by recent purchasing decisions on the Canary Wharf, clients and design teams do not restrict themselves to buying British, but purchase the components for the buildings on the world market. This is not new, but is perhaps an increasing trend as more experience is gained of the way in which overseas specialist contractors respond to the challenges. The on-site performance may not be any different, but the initial capability and input to the design process is different and very attractive to UK design teams. The system in which the overseas companies have developed demands the ability to input into the design a high quality engineering based 'design' input, which considers the problem in the context of the total project need. This ability is what the UK specialists must develop.

13

A blue print for sub-contracting

Sub-contracting is here to stay. It will take an increasing proportion of the total construction workload as more contractors see it as a way of surviving the volatile ups and downs of the construction market. From a sector of the industry which in the past has somehow lacked the respectability of general contracting it is now recognised that sub-contractors play a major role in the industry. The successful ones are increasingly specialized, are large and influential, and can control their terms of business. There will always be a high turnover of small companies as they will be used as a buffer against the volatilities of the varying workload. This is not unhealthy, it is a necessity if the industry is to survive. As the major sub-contractors flourish they in turn will use smaller sub-contractors and it is the interest of these two groups that we address here.

A healthy sub-contracting sector will be composed of successful larger companies, providing a comprehensive, high quality service. These companies need to be supported by a whole series of other sub-contractors who are striving to meet the quality objectives set by the major sub-contractors so that they in turn will become recognised as principal sub-contractors in their own right. The competition will be intense and it will be those sub-contractors who accept that they will have to rethink their business strategy who will survive and prosper. Much of the attraction of the foreign competition entering the UK is that these companies provide a highly engineered, quality product, efficiently and economically. They are familiar with dealing with designers and most of their managers are qualified engineers with the authority to make decisions for the whole business. The pressure is on the UK sub-contractors to behave similarly.

Some are rising to the challenge but the majority must follow, if sub-contractors are to take full advantage of the opportunities in the UK and Europe.

The strategy for the future is largely in the hands of the sub-contractors, but only if contractors, designers and clients use them effectively will their newly developed capabilities be used to the benefit of all. Here we have developed a two part strategy, the first is a longer term strategy for the next, say five years, which focusses on the action necessary to develop the high quality specialist that the industry needs. The second part is an action plan for immediate implementation which every sub-contractor can use to develop their company to meet the immediate business pressures.

A medium term strategy for sub-contractors

The strategic aim of any business is to economically produce a quality product, at a profit. This is of benefit to both the sub-contractor's business and the building project, the latter being required to support the former. The business must, therefore, be geared to providing a service to the project. It is this single aim which has been the most fundamental realization to many successful companies working on the major management projects over the last few years. The majority of sub-contractors will have to rethink their business strategy by examining every aspect of the business to see whether it is project orientated. Where it is not change must occur. Equally by removing the barriers to higher productivity, added value can be achieved and a competitive advantage gained.

At whatever point the sub-contractor joins the design, manufacture and supply chain, the individual parts of the operation must be focussed on the needs of the project. This will be in many cases be a fundamental about turn for many organizations that have developed in line with the more traditional, internally focussed, British attitude to business. To help in understanding how the business should develop each aspect of the design, manufacturing and assembly process will be examined to demonstrate the sort of approach that will be needed.

Design

Design teams are looking to the sub-contractors to make a significant contribution into specific aspects of the design process. They are looking initially for an ability to offer input at the conceptual stage which requires a visionary contribution founded on practicalities, but nonetheless able to explore the full flexibilities and beyond of the sub-contractor's capabilities. They are looking for a high grade intellectual input to match their own visions. This is why the European and American 'engineering' based sub-contract designers are so attractive in the early design stages, because they have empathy with the broad design objectives.

The second aspect that design teams are looking for now is the maximization of the site assembly process so that a high rate of production can be achieved on site. The sub-contractor's designer must, therefore, be able to conduct the 'trade off' between product, manufacture and site assembly, to achieve this efficiency, whilst still maintaining the design objectives. This requires that the designer is closely integrated with the site experience of the sub-contractor's assembly teams and that they are both striving to improve every aspect of both of their operations.

The third aspect is the introduction of project based CAD systems to which everyone will be required to input their design. Whilst this has been talked about for years the technology has now caught up with the ambition and is becoming a practicality. The promise is that the complexity of the design process will be simplified and speeded up with electronic transfer of data. The problem is that those who have invested in CAD will have to now invest in communication and translation software, whilst newcomers will have to ensure that they have compatibility with the latest CAD systems. Once again care will have to be exercised to ensure that decisions based on purely internal criteria do not cut off the wider, project based, opportunities.

To achieve the above aims, which will become increasingly vital to the design teams, the sub-contractors must make some basic decisions about the extent of their involvement before a contractual commitment. Some projects recognise this dilemma and use two stage bidding, but until the full costs of this 'free' design input are revealed it will be very difficult to resist an ever increasing demand to supply a major input to the early design. The time may be coming when the generosity implicit in this loss leader approach is curtailed when it is not paid for. As shown earlier it is essential to the design process, but hidden, because it is part of the 'sub-contract works'. Only when it's full extent is revealed to all members of the design team will it's integration be properly managed and many of the claims for delay stop.

Manufacturing

The manufacturing processes are long and complex but, they are only a means to an end and not an end in themselves. The trend is towards increasingly complex components and away from standardization. The percentage of 'one-offs' will rise to be the majority, thus destroying conventional production economics.

There are considerable added value advantages if individually tailored components can be supplied, but only if it is economic. Conventionally this was done through using time as a buffer to even out the process, but where time is being continuously squeezed this is no longer a viable option. Investment in flexible manufacturing systems must be made to allow the automation of the manufacturing process so that the range of components can be supplied efficiently. It will not be long before the designer's CAD system will interface to the manufacturing system thus opening up the ability to tailor the product to the actual need very much more quickly. In this way the flow of components required by the site can be supplied in the right order and so the erection process can be done efficiently.

Delivery

Many of the most efficient projects have used intermediate storage warehouses off-site to accept deliveries from the manufacturer which are repackaged into composite lots to then be delivered to the site. The implication is that the supply process is dictated by the factory. Once again the emphasis is wrong. The factory has the handling and packaging technology and it should not be left to six labourers and a forklift in a shed trying to reassemble loads for site delivery, with the consequent high liability to damage. The factory should be where the loads are scheduled and organized properly. Admittedly the manufacturers are in a powerful position to dictate terms, but they are only a service function and not the end product. They must be educated to supply in the order and at the rate needed.

Assembly

Site productivity must rise. There are not going to be enough people soon to build buildings if the current rates of productivity are maintained. There is scope to double the current productivity levels by organizing the task properly in the first place through the design and then by improving the site management. It is no good blaming others for the interruptions and disruption to the work. The sub-contractor should have good quality managers on site organizing their work, in conjunction with other contractors where necessary, to ensure maximum production. It is no longer acceptable to rely on the main contractor for the detailed site management. The management of the sub-contract tasks must be by the sub-contractor. This in turn requires that the sub-contractor employs and trains competent managers. The management task is to ensure that the whole design supply and construction process is flowing to meet the demands of the site assembly process. The site manager must have authority over the on and off-site work.

Only when the management have organized the tasks, material and plant can the pressure be applied to the operatives to perform. The operatives must understand the project objectives and their role in completing the work to the highest possible standard, there should now be no barriers in their way. They must respect the work of others.

The operative must also be respected and treated properly. Good pay, fair conditions, and training to upgrade their skills will all help to develop an *esprit decorps* and so foster the will to do better work.

Actions for the immediate development of sub-contracting

There are many issues which all sub-contractors must consider very carefully in addition to the longer term strategic issues if their business is to expand and grow. None of the actions are new, it is just that not many businesses apply them consistently. Those that do are seen as the successful companies and are the ones which clients, designers and contractors like working with repeatedly. The following are a number of questions to which the answer should be "yes". If it is not, then action is needed.

Does the company have a 'can do' attitude?

Are problems to be overcome or used as excuses for poor performance? A 'can do' attitude to problem solving is about accepting that those who identify the problem get on and solve it. Problems are there to be solved quickly and with no fuss. The implication is that people have to take responsibility for their actions and that decision-making happens at a very low level in the organization. Authority must also be given with responsibility, which implies that the level of trust throughout the organization rises.

Are the project objectives understood by everyone?

Communication throughout the company is vital. It is essential to focus everyone onto the key objective - satisfying the project need. What must be got over to everyone is that each project is not just another job, it is extremely important to those involved in the project. The company objectives are assumed to be known, but the project objectives are rarely known. Formal communication at all levels should, therefore, be undertaken, in the form of face to face discussion between the key people:

Current Trends

Away from nomination of sub-contractors and suppliers by the architect.

Towards greater specialisation and product specific skills, from specialist to 'super specialist'.

Towards shorter periods for training and apprenticeship.

Towards more self-employment in the industry.

More off-site pre-fabrication of components with a reduction in the manpower needed for assembly on site.

Greater awareness of the need to provide good quality at the right price with preference being given to quality assured products and firms, and making quality assurance mandatory on public sector projects.

More vertical integration of companies with specialist contractors providing an integrated design, manufacture, assembly and maintenance service for their craft or skill.

Clients becoming more involved in their projects and seeking longer guarantees and more collateral warranties.

Towards more litigation.

Towards passing the risks to another party by onerous contract conditions.

A growth in the importance of the speed of construction and the need to meet agreed completion dates.

More designs and build projects with architects and contractors becoming more aware of design and the environment.

More involvement of overseas clients, architects and contractors, particularly from the European Community, the USA, Japan and Korea.

More privately financed infrastructure projects, with contractors and specialists becoming involved in the financing arrangements.

Contractor to Directors

It should be made clear what is expected of the sub-contractor; what the difficulties are, where the problems are, where the limitations are and the detail of the interfaces with other trades. The Director is in a position to commit the company as well as communicate his commitments to his own people. This should be reinforced at each stage where the team of people involved changes significantly.

Director to key people

Formal discussion with all the key managers and people involved in the project, from design through manufacturing to the site operations is required, first to establish the needs of the project and then to review and reinforce them at regular intervals. This is not progress reviewing in the conventional sense, but a reinforcement of the commitment to achieve the objectives.

Site management to design and manufacture

There are often gaps in the understanding between these two groups about what the site is trying to achieve. Recent examples show that companies have used the client's marketing video to explain the project to the off-site people with dramatic effect. There has been an instant rapport with the site team and previous difficulties have been overcome with the new understanding. This can be achieved in other ways, but because the off-site people rarely see the site and, therefore, the problems, it is difficult for them to become committed to the project.

Foremen to workforce

Why is safety still a major problem? It is because it is rarely talked about - the operative is meant to know about the dangers. In practice the message must be hammered home daily. This amongst other things is the role of the 'tool box' meeting. Whilst the Japanese are the most widely publicised practitioners of this communication method it is gaining favour in other countries, even in the UK. Two trade contractors who have just returned from Japan are introducing it, as is one of the major contractors. It is common sense, in that an informed worker is a better worker. The foreman takes five minutes at the start of each shift to explain the day's objectives, work areas, safety issues and dangers, yesterday's problems and how they were sorted out, together with the more general issues. It reinforces the knowledge that the management cares about what is going on and is there to sort out the problems so that the workers can work effectively.

Is the latest technique and technology being used?

The question is, "How does the company know it is using the latest available techniques?" Unless it is actively seeking information on what its' competitors are doing or what is happening overseas in their field then the answer must be "No". Setting up low cost study trips to areas of known expertise is the first step. Borrowing ideas and adapting them is another step to picking up ideas which can be used to develop the existing practice. The aim of the development should be to increase productivity and provide a better quality product. The growth of off site pre-assembly is one way in which new production ideas, taken from other industries, has been used to achieve significant improvements in the overall quality and efficiency of the firm. Only when the technology within the product has been refined can the workforce be expected to perform effectively.

Is training being taken seriously?

This is an area of great difficulty because of the number of small firms and the fluidity of the business, and also the wide range of training agencies offering courses. The first thing is to look at the training need:

❑   The core workforce - update skills/techniques and develop multi-skill capability.

❑   New workers - ensure that there is a well paid career path with a recognisable pattern of skill acquisition.

❑   Management - offer an exciting career which demands high levels of skill and the ability to take on wide ranging responsibility.

❑   The entire workforce - the target is to aim for is retraining every two years to keep pace with the changes inside and outside the company.

Is the company actively working with the training agencies?

It is no good complaining about the lack of trained manpower if the company is not actively involving itself in establishing the requirements for its particular trade. The unions, Government, government agencies, trade associations and employers are all wrestling with the problem of providing a well trained workforce. However, it must meet the actual need and this is where action is needed to make sure that the objectives of the new training schemes, will produce the workforce that is needed. Each company should have defined it's needs and made sure that the agencies are aware of their specific needs. Where there is a short fall the company should consider how it is going to make it's own provisions.

Is developing a working relationship important?

This is really a question about attitudes to contracts and the implications of the various clauses that are written into them. Without exception the successful projects are run with the contract firmly shut in the desk drawer. Every attempt is made to resolve a problem before it occurs through developing sound working relationships based on trust and an ability to perform. Obviously the requirements in the contract must be understood and the risks properly evaluated, but in terms of the working relationships, the intention in the contract is what should be used.

Are onerous clauses a major worry?

They should not be if the organization understands the implications. Firstly the risk should be assessed and priced fully and secondly all the people in the organization should understand the implications of the contracts entered into. An action by someone in the organization, who is not aware of the implications of their action, can cause significant difficulties. These clauses are not a great difficulty if everyone is doing their job properly and sensibly.

Is liability for errors in design and subsequent latent defects a major issue?

Because specialist contractors are being increasingly requested to provide warranties of their performance this is an area of concern to everyone. Whilst most risks can be insured against it is important to ensure that all the risks are covered and that there are no gaps between the cover in the various policies. Expert advice is well worth the investment.

Desirable trends for the future

Away from confrontation towards co-operation.

Towards a bigger environmental lobby that will demand cleaner, safer systems of work that do not impact the environment.

More training of women for craft and operative careers and more training of people in mid-career who want to enter the construction industry.

Towards more single point responsibility with one party being responsible for design, construction, and maintenance.

Towards greater awareness and more responsibility for safety shown by the specialist contractors.

Recognition that risks should be identified, analysed, managed and discussed openly and constructively rather than transferring by tough conditions of contract.

Towards a better image for construction as an industry with good wages and better site conditions.

More joint ventures between contractors and specialist contractors recognising the ownership advantage of each partner.

A dramatic shift in the culture away from contractual claims that has inflicted the industry by involving contractors much more in the planning and design stage.

Greater European awareness and less of the 'European unity will never happen' attitude.

A growth of investment in training with sites holding induction training courses to explain to the craftsmen and operatives what is expected of them and make the workforce feel that they have something to contribute.

Recognise the merits of tried and tested standard forms of contract and using highly skillled specialised lawyers who deal solely with construction.

Are customer problems seen as business opportunities?

With the developing sophistication of knowledge within the specialist sub-contracting organization there is bound to be an increase in their involvement at the early design stage. Because of the leadership of the design process by the design team a major amount of this involvement will be to modify and develop the specialist's products and services to the specific project needs. Whilst there are obvious limits there are also opportunities to develop the business. Frequently this externally led innovation is looked upon as an irrelevant nuisance and the opportunity turned down, only to be done later by a competitor or a foreign company. This reluctance to grasp opportunities has been one reason for the growth in import penetration by overseas companies into the UK.

Is 1992 a threat or an opportunity?

Nearly every trade association has a committee studying the implications of 1992 and so there is a significant awareness of the impending harmonization of the market in the European Community in 1992. There are three issues. The first is the effect of harmonization of regulation and standards. This has been the major issue to be addressed by these committees and every trade should be aware of what is being decided and how it will affect their business. Secondly, is the likely threat of massive inroads into the UK market by European contractors. At the moment this does not seem to be a worry, but as has been seen over the past few years there has been a steady influx of foreign specialist contractors, who have established bases in the UK as a precursor to 1992. With the buoyancy of the UK market this is likely to continue. Thirdly, is the opportunity for exports to the EC. At the moment companies appear to be still making up their minds about this.

Does your business contribute to the import deficit?

At the moment there is an imbalance in construction imports over exports of over £3 billion. There are massive opportunities to reduce this by increasing the UK manufactured content of products. The European concept of joint venturing to increase the capabilities of the single organization are worth exploring here to share the risk as well as the opportunity.

Setting the conditions for effective sub-contracting operations

The above developments are practical steps in achieving a quality, project orientated, sub-contracting organization capable of responding to the demands from; clients, designers and contractors. As sub-contractors work in response to the business and project environment generated by the participants they in turn must respect the sub-contractor's achievement potential, if given the right working conditions. They must create the right working conditions.

Contractors

A new working relationship is needed. As more risk is passed to the sub-contractor the contractor is at greater risk if the sub-contractors fail, it is, therefore, in the interest of the contractor to maintain all his sub-contractors at a high level of efficiency and thus ability to complete his work. Sub-contractors are in a difficult enough situation already, having to answer to both designers and contractors; increased difficulties through poorly drafted contracts and poorly managed payments will only exacerbate their situation. The contractors must become sympathetic to the needs of the sub-contractors. If the sub-contractor cannot achieve his targets it is up to the contractor, in his management capacity to find out what is wrong and help the sub-contractor to solve the problem. This is not a case for taking over the sub-contractor's operations but a case for checking that, at every stage of the process in which the sub-contractor is involved, the contractor has ensured that the sub-contractor has all the information and resources that he needs. Only when this has been done can the contractor criticize the sub-contractor. Construction management has changed dramatically in the last five years with the increase in sub-contracting and the role that sub-contractors are now asked to perform. The contractor's role is essentially a mothering and supportive role to enable the sub-contractor to perform.

Attitudes within contracting must change. The whole organization, both on and of site must take responsibility for avoiding failure. When the complex system of design and construction fails, the contractor has failed, because he has not foreseen the possibility of failure and taken prompt action. The management must be constantly thinking ahead and be involved in every aspect of the sub-contracting process to ensure that it will not fail. Setting broad based targets and writing tough contracts does not work. Only by monitoring events in a detailed way against a good knowledge of what should be done will the required be achieved.

Contracts must be drawn up which truly reflect the needs of the project and are sympathetic to the realities of the sub-contracting task. As demonstrated elsewhere in this study, attempts to manipulate the construction process through forms of contract often fail because the true roles and responsibilities that must be in place for a project to succeed are not recognised. Once in place, the contract must be managed fairly. Project success requires financially healthy sub-contractors. Variations must be valued promptly and incorporated in the other payments which must also be made promptly and correctly. In too many cases junior surveyors gain their spurs by, at best, gamesmanship. Often this is done without the knowledge of senior management who are trying to develop a sensible working relationship only for it to be undermined.

Many contractors are trying unsuccessfully to reduce their financial and managerial risks. In fact, with the increase in management only contractors and contracts, the contractor's management must rapidly recognise that it has a new role to play, that of 'facilitator', to ensure that the sub-contractors can perform. It is a much more difficult role, than they have traditionally undertaken, and requires much greater involvement.

Design teams

Specialist sub-contractors now make a significant contribution to the design process and design teams cannot function without it, yet they seem increasingly reluctant to recognise this. The design process is very complex and a large proportion of a project's drawings are originated by specialists. It is particularly at the detailed level where this input is made and the design team must organize itself accordingly. Designers are increasingly becoming co-ordinators of input and yet few of them have any useful project oriented management systems. The consequences of this are that the sub-contractor's input is not organized as to timing and the information supply process fails. Design teams must start managing the production of detailed design information properly. This necessitates introducing systems that can handle complex detail in the same way that contractors plan and manage the complexities of construction.

Because the costs of the sub-contractors design process are not revealed the costs of failure are only apparent when the site activity is disrupted. This raises the fundamental question of the value of the sub-contractor's knowledge. It is not without value, but historically it has been given freely in the hope of attracting business. However, it is time to rethink this approach and for the design team to clearly define the knowledge they need and where appropriate pay for it. This is particularly true of the pre-competitive situation where 'advice' on methods, detail and capability is being sought.

The contracting systems in nearly all other countries are designed to harness the full productive capability of the sub-contractor by carefully defining the boundaries and interfaces of the package of work so that the potential for interference is minimized. In the UK the sub-contractor is seen as the facilitator of the design and thus subservient to the design team. These two views must be reconciled if the problem of the low level of productivity endemic in the UK approach is to be resolved. Designers must allow the sub-contractors freedom within their speciality to maximize both the aesthetics of the design and the production processes. It would not necessarily mean a loss of sovereignty, if intelligently managed.

Conclusion

The sub-contracting industry covers such a width of interests and organizations that a general study of this nature cannot answer the particular problems or discuss the specific issues that everyone would want. The study could be read as one which is extremely critical of the industry, but this is not so. The intention is to raise the issues and to explain why they present the problems that they do. The explanations may be uncomfortable to some, but are necessary if progress is to be made. There are many difficulties ahead. The uncertainties in the industry will remain, as will the volatility of demand. The effect of the EC harmonization is largely unknown and is adding to the difficulty of predicting the future. However, it is also clear that there are significant opportunities to those companies who are going to invest in the future. Growth will come for those who develop their specialisation within an organization committed to quality and service.

Bibliography

Ball M, (1988) Rebuilding Construction, Routledge, London

Bennett and Ferry (1988), Specialist Contracting - A Review of the Issues Raised in Their New Role in Building - a report for the Building Centre Trust, Centre for Strategic Studies in Construction, Reading University (unpublished)

Blackspot Construction (1988), Health and Safety Executive, HMSO , London

Brickies add 15% to cover pay when paid, Contract Journal, 27 April 1989

Electrical contractors to withdraw from CITB, Building, p7, 19 February 1989

FASS list slams slow contractors, Building, p7, 25 September 1987

Faster Building for Commerce (1988) NEDO, Millbank, London

Gray C, (1981) Management of the Construction Process, Building Construction and Management, March Issue, CIOB Ascot

HSC and HSE Annual report, 1987/88, HMSO, London

How to take part in the Quality Revolution: A management guide. P A Consulting Group, Bowater House East, 68 Knightsbridge, London

Huxtable, (1988) Remedying Contractual Abuse in the Building Industry, Technical Information Service No 99, Chartered Institute of Building, Ascot

JO (QS) Survey of contracts in use, Chartered Quantity Surveyor , RICS, London, January 1989 p24-26

Jacques E, Gibson, R O, Isaac, D J, (1978) Levels of Abstraction in Logic and Human Action, Heinemann, London

NJCC warning on design warranties, Chartered Builder, October 1988

Passing the buck, Building Cladding and Curtain Walling supplement, Issue 16, No 7594

Pay and training are targets for revolutionary reform, Building p7, 19 February 1989

Peters T and Austin N, (1985) A Passion for Excellence, Random House, New York

Pinchot G, (1985), Intrapreneuring, Harper and Row, New York

Quarterly Bulletin of Housing and Construction Statistics, DoE, HMSO, London

10 Year Housing and Construction Statistics 1977-87, DoE, HMSO, London

Unwin Brothers Limited, The Gresham Press, Old Woking, Surrey, England